Introduction to
Thermoacoustic Devices

Introduction to
Thermoacoustic Devices

Tetsushi Biwa

Tohoku University, Japan

W♦ World Scientific

EW JERSEY · LONDON · SINGAPORE · BEIJING · SHANGHAI · HONG KONG · TAIPEI · CHENNAI · TOKYO

Published by

World Scientific Publishing Co. Inc.

27 Warren Street, Suite 401-402, Hackensack, NJ 07601, USA

Head office: 5 Toh Tuck Link, Singapore 596224

UK office: 57 Shelton Street, Covent Garden, London WC2H 9HE

Library of Congress Cataloging-in-Publication Data
Names: Biwa, Tetsushi, author.
Title: Introduction to thermoacoustic devices / Tetsushi Biwa, Tohoku University, Japan.
Description: Hackensack, NJ : World Scientific, 2022. |
    Includes bibliographical references and index.
Identifiers: LCCN 2021001910 | ISBN 9781944659769 (hardcover) |
    ISBN 9781944659776 (ebook for institutions) | ISBN 9781944659783 (ebook for individuals)
Subjects: LCSH: Heat exchangers. | Heat--Transmission. | Acoustic emission.
Classification: LCC TJ263 .B59 2022 | DDC 621.402/5--dc23
LC record available at https://lccn.loc.gov/2021001910

British Library Cataloguing-in-Publication Data
A catalogue record for this book is available from the British Library.

For any available supplementary material, please visit
https://www.worldscientific.com/worldscibooks/10.1142/Y0023#t=suppl

Typeset by Stallion Press
Email: enquiries@stallionpress.com

Preface

Sound waves propagating down a tube behave dramatically different from those in a free space, because the thermodynamic aspects of sounds are fully unwrapped by the thermal interactions of the gas with the tube wall. Generation of sound by heat, cooling effect by sound and enhancement of heat transport from hot to cold are examples of basic thermoacoustic phenomena, from which novel acoustic devices have been made possible such as *acoustic engine*, *acoustic cooler*, and *dream pipe*.

In relation to energy and environmental issues currently encountered in industry, thermoacoustic devices are gaining much attention owing to inherent advantages like simple structure, no/less moving solid parts for operation, and no use of gases with high global warming potential and Ozone depletion potential. Furthermore, thermoacoustic devices offer new topics although heat and sound are classical subjects of physics and engineering.

This book is intended to provide a practical guide for students and engineers who are interested in designing, building, and analyzing thermoacoustic devices and those who are fascinated by thermoacoustic phenomena out of curiosity. Many readers will find that undergraduate level knowledge of mechanical vibrations, acoustics, and mathematics are sufficient to go through the book.

The structure of this book is schematically illustrated in Fig. 1.

Chapter 1 introduces various examples of thermoacoustic devices to identify common elementary components. In particular, stacks and regenerators are indispensable components of thermoacoustic

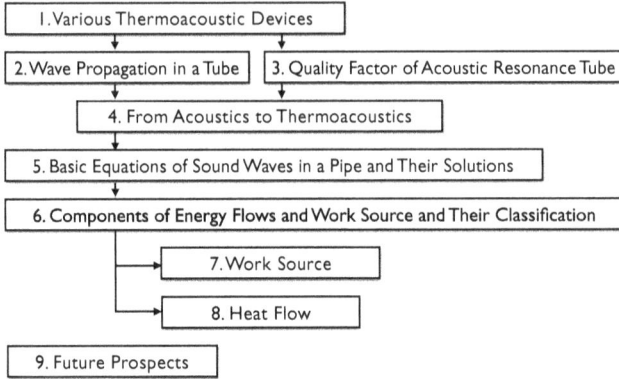

```
┌─────────────────────────────────────────┐
│  1. Various Thermoacoustic Devices        │
└─────────────────────────────────────────┘
   ↓                              ↓
┌──────────────────────┐ ┌───────────────────────────────────────┐
│ 2. Wave Propagation in a Tube │ │ 3. Quality Factor of Acoustic Resonance Tube │
└──────────────────────┘ └───────────────────────────────────────┘
   ↓
┌─────────────────────────────────────┐
│  4. From Acoustics to Thermoacoustics │
└─────────────────────────────────────┘
   ↓
┌───────────────────────────────────────────────────┐
│ 5. Basic Equations of Sound Waves in a Pipe and Their Solutions │
└───────────────────────────────────────────────────┘
   ↓
┌──────────────────────────────────────────────────────────────┐
│ 6. Components of Energy Flows and Work Source and Their Classification │
└──────────────────────────────────────────────────────────────┘
        ┌──────────────────────────┐
     →  │  7. Work Source           │
        └──────────────────────────┘
        ┌──────────────────────────┐
     →  │  8. Heat Flow             │
        └──────────────────────────┘
┌──────────────────────┐
│  9. Future Prospects   │
└──────────────────────┘
```

Figure 1: Structure of the book.

devices. To gain physical understanding of the role of the components, the following chapters are organized as follows. In Chapter 2, the problem of sound propagation in a tube is introduced. The propagation constant is governed by thermal and viscous interactions of the gas with the tube wall. A dimensionless parameter, $\omega \tau_\alpha$, showing the degree of interactions, consistently appears throughout the book. Chapter 3 discusses the quality factor of the acoustic resonance tube. When the thermally induced self-sustained acoustic oscillations are viewed from an energetic point of view, the quality factor becomes key to understand the occurrence of thermoacoustic spontaneous oscillations. Chapter 4 introduces a central physical concept of *heat flow* and *work flow*, which renews the schematic diagram of heat engines. Chapter 5 provides the entropy fluctuations of the working fluid. By introducing a Lagrangian point of view, it is shown that fluid parcels act as small-scale heat engines when acoustic gas oscillations take place under the influence of thermal interaction with the tube wall. Chapter 6 graphically documents the basic physics of energy flows and work source. The contents of this chapter is the main achievement of the thermoacoustic theory originally developed by Tominaga. Chapters 7 and 8 show working mechanisms of thermoacoustic devices based on the results of the thermoacoustic theory. Finally, Chapter 9 provides future prospects of thermoacoustic devices.

The topics in each chapter reflect my personal experience and interest. Readers are recommended to read the papers and textbooks listed in the Bibliography to gain perspective on a wider variety of topics as well as keep up-to-date on the latest articles, as thermoacoustics is a rapidly growing field of study.

I received lots of advice and encouragement from Prof. Taichi Yazaki during the preparation of this book. I was extremely fortunate to have the opportunity to work with him. I am also grateful to Prof. Akira Tominaga. I learned many things from him. I thank Dr. Tatsuo Inoue, Prof. Yuki Ueda, and Dr. Hiroaki Hyodo for inspiring discussions and helpful comments. I would also like to thank my laboratory members who carefully and repeatedly checked the equations and texts: Mr. Shun Tamura, Mr. Hayato Kaneko, Ms. Moeko Sato, Mr. Keita Muraoka, Mr. Yusuke Takayama, and Mr. Takayuki Kobayashi.

Sendai, Japan, 2020

Tetsushi BIWA

Contents

Chapter 1

Various Thermoacoustic Devices

We begin with a brief description of the history of thermoacoustics, followed by a classification of various thermoacoustic devices from viewpoints of their function and structure. Relevant heat engines are also introduced. The working mechanisms will be discussed in later chapters.

1.1 Brief History of Thermoacoustics

Figure 1.1 lists research subjects and achievements from classical thermoacoustic oscillations to modern thermoacoustic devices. It also includes the treatises and textbooks on thermoacoustics. The history of thermoacoustics will be described in this section.

1.1.1 *Dawn of thermoacoustics*

It took more than a quarter and a century for the problem of sound speed, originally posed by Newton in *Principia*, to be solved by Laplace. Newton assumed Boyle's law on gas density and pressure to determine the sound speed in air, and hence he postulated a constant gas temperature during sound propagation. His theoretical value was 20% less than the observation. To remedy the discrepancy, much effort was made by scientists including Euler and Lagrange, but to no avail. It was the thermodynamic consideration that led Laplace to find the correct value of sound speed. He assumed adiabatic processes of the gas, where the gas temperature goes up and down with acoustic pressure. The problem of sound speed can be

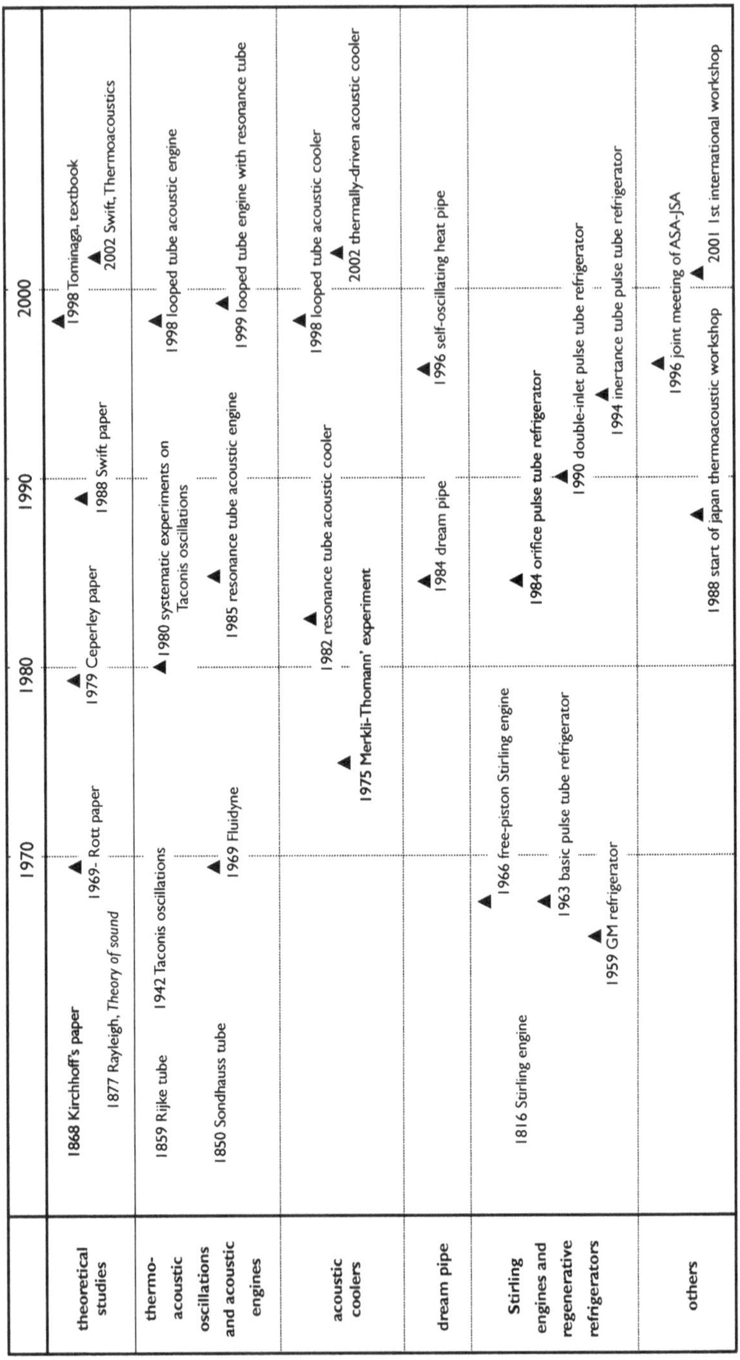

	1970	1980	1990	2000
theoretical studies	1868 Kirchhoff's paper 1877 Rayleigh, *Theory of sound* 1969- Rott paper	1979 Ceperley paper	1988 Swift paper	1998 Tominaga, textbook 2002 Swift, Thermoacoustics
thermo-acoustic oscillations and acoustic engines	1859 Rijke tube 1942 Taconis oscillations 1850 Sondhauss tube 1969 Fluidyne	1980 systematic experiments on Taconis oscillations 1985 resonance tube acoustic engine		1998 looped tube acoustic engine 1999 looped tube engine with resonance tube
acoustic coolers		1975 Merkli-Thomann' experiment	1982 resonance tube acoustic cooler	1998 looped tube acoustic cooler 2002 thermally-driven acoustic cooler
dream pipe			1984 dream pipe	1996 self-oscillating heat pipe
Stirling engines and regenerative refrigerators	1816 Stirling engine 1966 free-piston Stirling engine 1963 basic pulse tube refrigerator 1959 GM refrigerator		1984 orifice pulse tube refrigerator 1990 double-inlet pulse tube refrigerator 1994 inertance tube pulse tube refrigerator	1998 double-inlet pulse tube refrigerator
others			1988 start of japan thermoacoustic workshop	1996 joint meeting of ASA-JSA 2001 1st international workshop

Figure 1.1: History of thermoacoustics. Topics and subjects of this book are shown.

considered as a starting point in the history of thermoacoustics, since it demonstrates the importance of thermodynamics in solving an acoustics problem.

Various thermodynamic effects on sound propagation were already found in the 19th century. *The Theory of Sound*, published by Rayleigh [1] in 1877, describes Kirchhoff's discussion of thermoviscous effects on sound propagation in a tube, and thermally-induced oscillations of gas columns in Rijke tube, Sondhauss tube, and tubes with internal heat source of hydrogen flame. Rijke tube is an open-ended pipe that has a piece of fine metallic gauge fixed in the lower part of the tube [2]. When the gauge is heated by a gas flame placed under it, a sound of considerable intensity is observed soon after the removal of the flame. Sondhauss tube is a tube with one end open and the other end closed by a sphere bulb [3]. By externally heating up the bulb, the tube starts to emit a sound. The sound source can be a flame burning in the tube [4]. The problem of a *singing flame* is considered as the most simple example of combustion-driven oscillations observed in modern industry. All of these phenomena constitute the routes of current thermoacoustic studies.

1.1.2 *Rott's study*

At the beginning of the 20th century, it became possible to liquefy helium gas in cryogenic laboratories. Thermally induced gas oscillations were also found in pipes which were inserted into a liquid helium vessel from the room temperature side, indicating that the oscillations are excited not only by heating but also by cooling. When these oscillations appear, considerable heat transport takes place and gives rise to an abnormal evaporation of the liquid helium. This phenomenon was named Taconis oscillation after the researcher who reported it for the first time [5]. It was Rott that made elaborate theoretical analysis of Taconis oscillation as a stability problem of fluid dynamics [6–12]. His theory predicted quantitatively the critical conditions of Taconis oscillations, and also contributed to the analysis of thermoacoustic cooling effect observed by Merkli and Thomann [13].

1.1.3   *Studies conducted by Los Alamos group and Tsukuba group*

Wheatley and Swift of Los Alamos Laboratory and Tominaga and Yazaki of Tsukuba University made remarkable contributions for the application of thermally induced gas oscillations. Wheatley and Swift, inspired by oscillatory flow-based cooling mechanism of pulse tube refrigerators, succeeded in developing an acoustic cooler by extending the operation frequency to audible frequency range [14, 15]. Tominaga and Yazaki made systematic studies on Taconis oscillations that confirmed the validity of Rott's theory. Encouraged by successful application of Rott's theory, Tominaga tried to make a theoretical analysis of the energy conversion and heat transport mechanisms in acoustic engines and coolers, especially emphasizing the importance of temporal phase difference between pressure and velocity oscillations of a gas column. His theory was found to universally explain the operation principles of conventional oscillatory flow-based heat engines like Stirling engine, pulse tube refrigerators, and GM refrigerators. His formulation is also applicable to dream pipe, where heat transport from hot to cold is significantly enhanced by longitudinal oscillations of liquid columns. Now his theory [16], as well as Swift's book [17], has been accepted as a standard theoretical tool for understanding and analyzing thermoacoustic devices. Swift and Ward built a computer simulation code, DeltaEC, using the result of the thermoacoustic theory. This code has been used extensively by thermoacoustic researchers as a design tool of thermoacoustic devices. Development of acoustic engines demonstrating a very high thermal efficiency significantly owes its success to DeltaEC [18].

The most striking achievement by Wheatley and Tominaga would be a proposal of basic physical concept of *heat flow* and *work flow*, which highlights a thermodynamic aspect of thermally-induced gas oscillations. As we address in Chapter 4, heat flow and work flow are formulated for a periodically steady fluid flow by integrating the energy equation of the fluid over an oscillating period and cross section of a flow channel. By further integrating over an axial coordinate of the flow channel, the energy equation expressed by heat flow and work flow is transformed to the first law of thermodynamics representing a mutual conversion between heat and work. In other words, heat flow and work flow serve as a *ladder* to connect *microscopic* description of fluid dynamics and *macroscopic* description of thermodynamics. The direct observation of work flow was made by Yazaki and Tominaga for the first time through simultaneous measurements of pressure oscillation and velocity oscillation in thermally induced gas oscillations [19]. The indirect measurement method known as the two-sensor method is often adopted because of the ease of experiments [20–22]. The work flow measurements are recognized as an indispensable experimental technique for understanding of thermoacoustic devices.

1.1.4 *Current research trend*

Figure 1.2 presents a search result of journal papers that was obtained by using the keyword *"thermoacoustic"*. The annual number of journal papers started to increase rapidly after the 1990s, probably because the analysis based on heat flow and work flow helped in understanding and fabricating various thermoacoustic engines. In 1996, a joint meeting of the American Society of Acoustics and Japan Society of Acoustics was held in Hawaii, USA where exchanges

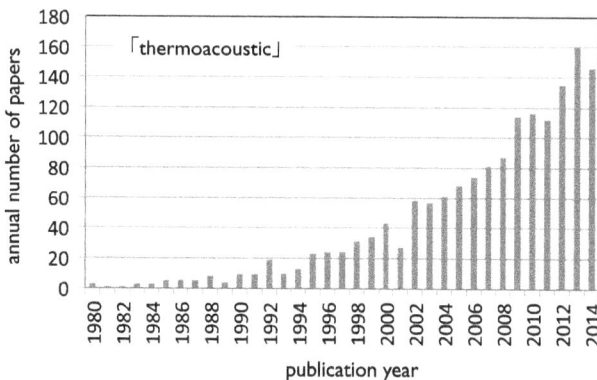

Figure 1.2: Annual number of papers on thermoacoustics.

of research achievements were done among researchers with different backgrounds. In 2001, the first international workshop of thermoacoustics was held in the Netherlands, and the second one in Sendai, Japan. Research works started in Los Alamos Laboratory and Tsukuba University have now grown a research community that spreads all over the world. The scope of research includes basic understanding of thermoacoustic phenomena and application such as developing acoustical heat engines. For a comprehensive list of articles on thermoacoustics, see Reference [23] by Garrett, who is also one of the pioneers.

1.2 Classification of Thermoacoustic Devices

A rich variety of thermoacoustic phenomena have been observed. When a steep temperature gradient is locally created in a gas column, longitudinal acoustic gas oscillations are induced. Such a self-sustained oscillation is called thermoacoustic oscillation. When a gas column is forced to oscillate by an external sound source like a loudspeaker, a part of the gas column is cooled down. Also, an oscillatory flow greatly enhances heat transport from hot to cold. These phenomena constitute the basis of thermoacoustic devices known as acoustic engine, acoustic cooler, and dream pipe, as summarized in Fig. 1.3. The acoustic engine is a heat engine that generates output acoustic power using thermoacoustic oscillations; the acoustic cooler is a kind of heat pump whose input power is

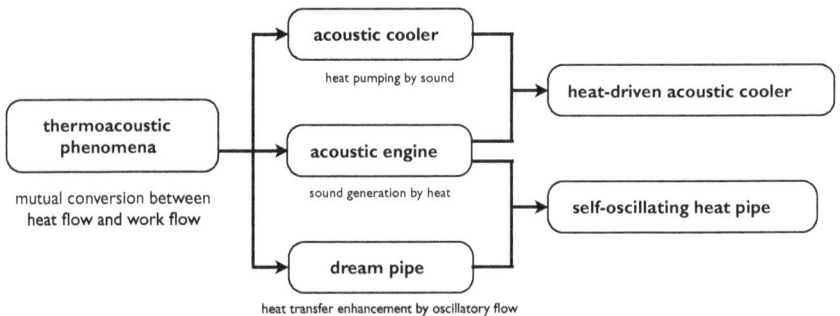

Figure 1.3: Classification of thermoacoustic devices.

maintained by sound waves; and the dream pipe is an oscillating flow-based heat transport device.

It is possible to combine thermoacoustic devices of different functions to create new ones. When an acoustic engine is connected to an acoustic cooler, a thermally-driven acoustic cooler can be built, where the cooler is energetically sourced by the output acoustic power of an acoustic engine. A self-oscillating heat pipe [24] would be considered as a combination of an acoustic engine and a dream pipe; it transports heat from hot to cold automatically when a temperature difference between the ends exceeds a critical value.

A more general classification becomes possible if heat flow and work flow are used as central concepts. As we will show in Chapter 4, the acoustic engine converts heat flow to work flow, whereas the acoustic cooler promotes heat transport from cold to hot thanks to the energy conversion from work flow to heat flow. In this section, we show some examples of thermoacoustic devices and relevant conventional heat engines.

1.3   Acoustic Engine

1.3.1   *Thermoacoustic self-sustained oscillation*

A primitive prototype of the acoustic engine can be constructed by a simple setup shown in Fig. 1.4. An empty tube, containing a *stack* and a pair of heat exchangers, forms an acoustic resonance tube with one end open and the other end closed by a plug. The stack used here is a ceramic honeycomb catalyst support used in automobile engines, but can be replaced with a stack of thin plates. In reality, the stack was named after such a structure. The working gas is air. The stack is placed in the middle of the resonance tube. As we will show in Chapter 7, the stack position is an important parameter to generate acoustic oscillations in the resonance tube with a relatively small temperature difference.

By heating up the heat exchanger near the closed end with a hand burner, a rather loud sound sets in after around 30 seconds. As noted by Sondhauss, a few droplets of water in the stack drastically shortens

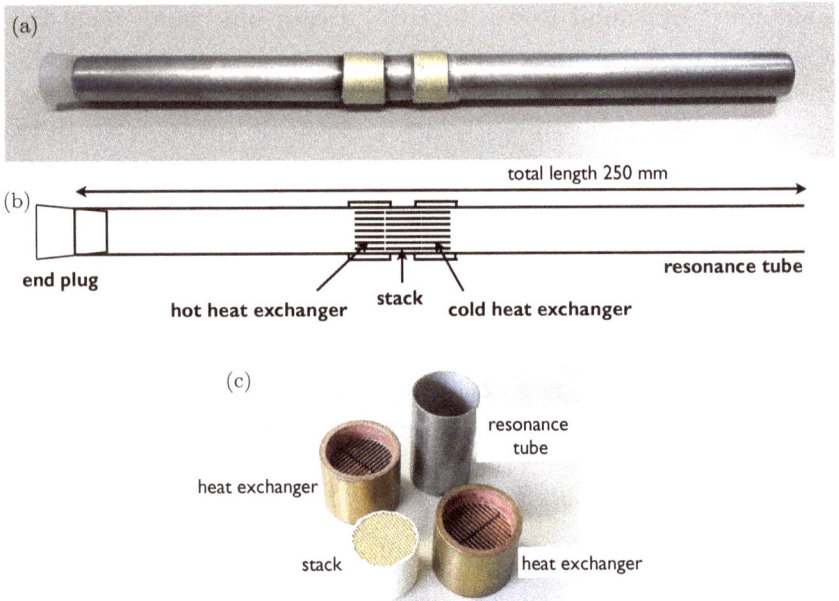

Figure 1.4: Thermoacoustic demonstrator (resonance tube acoustic engine). (a) photo, (b) scale drawing, (c) photos of stack, heat exchanger, and a part of resonance tube.

the heating time. The sound frequency is around 300 Hz, reflecting a fundamental longitudinal oscillation mode of the resonance tube.

The pipe length and diameter are 250 mm and 15 mm, respectively, in the setup shown in Fig. 1.4. The larger one with 600 mm length and 40 mm diameter is also capable of generating sound. The sound frequency is reduced as the pipe length is extended. Reference [15] also describes the demonstrator of the resonance tube acoustic engine.

The phenomenon of thermoacoustic oscillations can be thought of as a kind of heat engine. Whereas a conventional heat engine like an internal combustion engine produces shaft power through periodic motion of solid pistons, an acoustic heat engine generates acoustic power through oscillating motion of gas. Indeed, the acoustic power production in the stack region was evidenced by simultaneous measurements of pressure and velocity in the resonance tube acoustic engine [19].

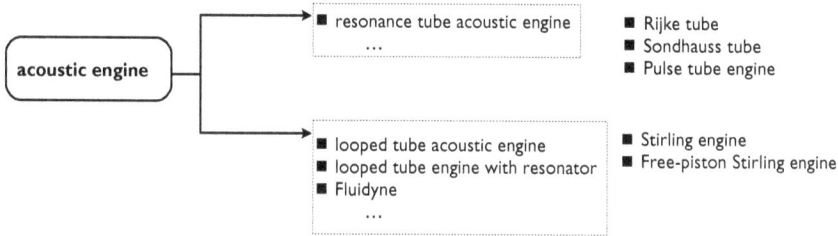

Figure 1.5: Classification of acoustic engines.

The acoustic engine is classified into two types as shown in Fig. 1.5: one is the resonance tube engine made of a resonance tube with open and/or closed ends, and the other one is the looped tube engine having a looped tube. In the next section, we will show examples of both types of acoustic engines.

1.3.2 *Resonance tube acoustic engine*

The resonance tube acoustic engine operates with the natural oscillation mode of a gas column in the resonance tube. The oscillation mode of the engine shown in Fig. 1.4 is the fundamental one with pressure antinode at the closed end and pressure node at the open end, where the tube length is essentially equal to the quarter wavelength. In a closed-closed resonator, the half-wavelength mode is likely induced in the engine. Rijke tube and Sondhauss tube fall into this category, as they are made of tubes with both ends open and the tubes with open and closed ends. The resonance tube acoustic engine is also called a *standing wave engine*, because the acoustic oscillations of the standing wave mode contributes to the energy conversion from heat flow to work flow, as described in later sections in detail. In short, the standing wave mode oscillation means that a fluid parcel oscillates with the pressure oscillations in phase with the displacement oscillations.

A combination of a liquid column and a gas column is also used to build the resonance tube acoustic engine, where the natural oscillation mode usually has very low frequency compared to the engine made only of a gas column due to the higher density of liquids than gases. A put-put toy boat with a boiler and a tube is an example

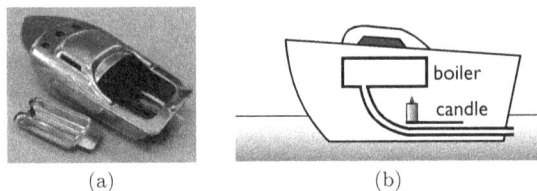

Figure 1.6: Put-put toy boat: (a) photo and (b) schematic.

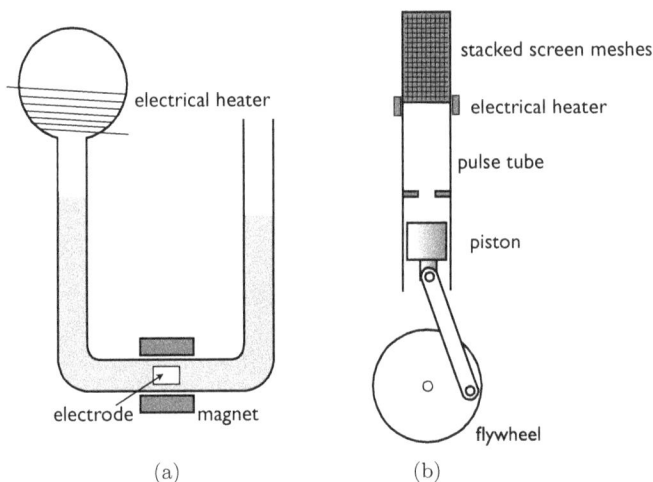

Figure 1.7: Resonance tube engine using natural oscillations of liquid column in a U tube (a) and pulse tube engine having a piston–flywheel assembly (b).

of such an engine (see Fig. 1.6). Another example is the engine made by Heulsz [25], aiming at Magneto-Hydrodynamics electricity generation system [Fig. 1.7(a)]. The basic structure resembles the Sondhauss tube, but a current-conducting liquid occupies the U-tube section in the middle. Heating up the bulb part generates the oscillations with frequency of 5.3 Hz. A pair of permanent magnets are aligned in a way that the liquid oscillates perpendicularly through the magnetic field. In this way, ac electric current can be extracted from the electrodes when the self-sustained oscillations are excited.

If one connects a piston with a flywheel at the end of a resonance tube, one can build a pulse tube engine shown in Fig. 1.7(b). In this engine, the acoustic amplitude is determined by a piston stroke, but the oscillation frequency speeds up with increasing heat input. Thus, it may not be appropriate to put this engine into a resonance tube engine category although it is one of the standing wave engines, since the measurements of pressure and velocity reveal that the standing wave mode oscillations of a gas contributes to energy conversion [26]. Owing to the flywheel rotation dynamics, the operation frequency is rather low compared to a gas-based resonance tube engine.

1.3.3 *Looped tube acoustic engine*

The looped tube acoustic engine was built in 1998, quite recently compared to the resonance tube acoustic engine. In the looped tube engine in Fig. 1.8(a), the oscillation mode is determined by a periodic boundary condition. Among the possible oscillation modes, the traveling wave mode oscillations of a gas was observed to contribute to energy conversion [27]. Therefore, the looped tube engine is also called a *traveling wave engine*. As we explain in more detail in Chapter 6, the traveling wave mode oscillation means that a fluid parcel oscillates with the pressure oscillations in phase with the velocity oscillations.

Connecting a branch tube improves the looped tube engine performance [28]. As shown in Fig. 1.8(b), the branch tube is rather large compared to the size of the loop. Reflecting the branch resonator, the oscillation mode looks like a standing wave mode oscillation as a

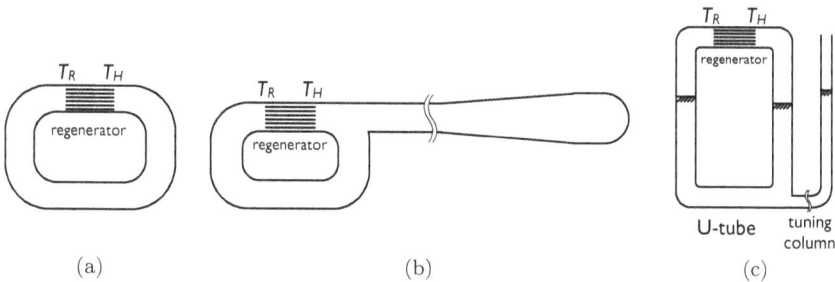

Figure 1.8: Looped tube acoustic engine. (a) looped tube, (b) looped tube with resonance tube, (c) Fluidyne. T_R and T_H denote the regenerator's ambient and hot ends, respectively.

whole, but the gas parcel in the regenerator region locally undergoes pressure and velocity oscillations similar to traveling wave mode oscillations. Therefore, this engine also belongs to a traveling wave engine. The acoustic power flowing out of the loop to the branch tube was 700 W, with a thermal efficiency of 30%, corresponding to 40% of the Carnot efficiency [28, 29]. Such a high efficiency is comparable to that of conventional heat engines such as an internal combustion engine.

Fluidyne, or a liquid piston Stirling engine, is one of looped tube engine, where a liquid column in U tube is used as a part of the loop, as shown in Fig. 1.8(c). When a sufficiently high temperature difference is given to a regenerator, the liquid column starts to oscillate with a frequency of around 1 Hz. West [30] introduced various types of such engines in his book, and described that *in the author's experience, it is quite difficult to construct a small Fluydine that cannot be made to work.*

Whereas a stack is used in the resonance tube acoustic engine, a regenerator is employed as the heart of the engine in the looped tube acoustic engine. The regenerator is often constructed by a dense stack of mesh screens, resulting in much narrower flow channels than the stack. Therefore, the thermal interaction of the gas with the channel wall is more enhanced in the regenerator than in the stack. A good thermal efficiency of a traveling wave acoustic engine originates from a good thermal contact attained in the regenerator.

The regenerator is also an essential component of a Stirling engine, which was invented in the 19th century. The Stirling cycle, being the thermodynamic cycle in the Stirling engine, is one of thermally reversible cycles operating between two heat baths. Therefore, it can in principle achieve a Carnot efficiency, but a complicated mechanical structure obstructed the wide application of the engine. Aiming at simplifying the structure, a variety of Stirling engines have been proposed as shown in Fig. 1.9(a)–(d). It was Ceperley that pointed out the similarity between the Stirling engine and the looped tube engine [31]. Chapter 4 and the later chapters will explain his idea in more detail.

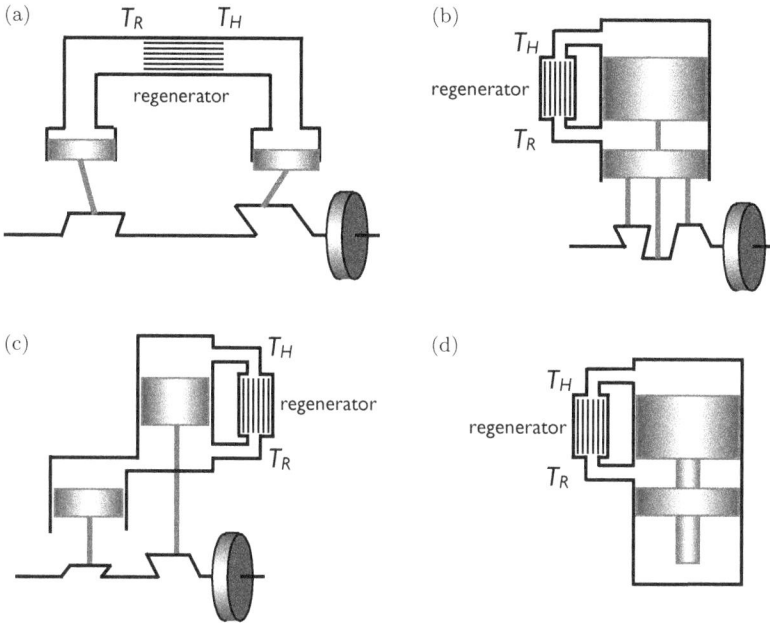

Figure 1.9: Basic types of Stirling engines: (a) α-type, (b) β-type, (c) γ-type, and (d) free piston type.

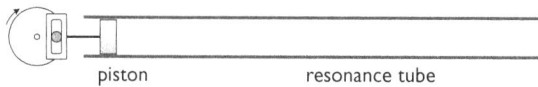

Figure 1.10: Experimental setup of Merkli-Thomann's experiment.

1.4 Acoustic Cooler

Merkli and Thomann theoretically predicted a cooling effect by sound when a gas column confined in a resonance tube was forced to oscillate, on the basis of Rott's theory [13]. They observed that the temperature at the velocity amplitude maximum of the resonance tube was lowered because of the acoustic oscillations of the gas column in the setup shown in Fig. 1.10. Their experiment can be considered as an early demonstration of an acoustic cooler.

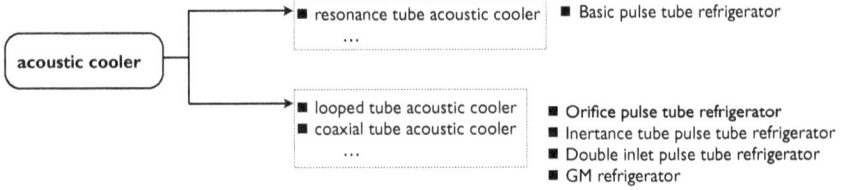

Figure 1.11: Classification of acoustic coolers.

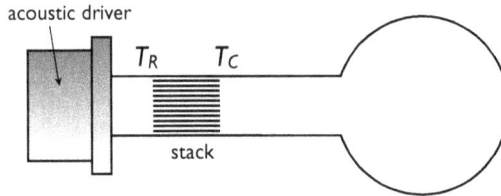

Figure 1.12: Resonance tube acoustic cooler.

Various acoustic coolers have been developed since the 1980s. They can be classified into two types depending on the structure as presented in Fig. 1.11. One is a resonance tube cooler installed with a stack, and the other is a looped tube acoustic cooler having a regenerator in place of the stack. Since pressure oscillation and velocity oscillation of a working gas in a regenerator is essential to the cooling mechanism of a regenerative refrigerator, it is also considered as an acoustic cooler. Some examples of these acoustic coolers are given in the following.

1.4.1 *Resonance tube acoustic cooler*

Wheatley and his colleagues developed a resonance tube acoustic cooler, motivated by the basic pulse tube refrigerator, which produces low temperatures by pulsating gas flow introduced in an empty tube called a pulse tube. Figure 1.12 presents a schematic of the resonance tube acoustic cooler, made of a Helmholtz resonator and a stack [32]. The resonator was filled with helium gas of 1 MPa, and was driven by an acoustic driver operating with the natural frequency of

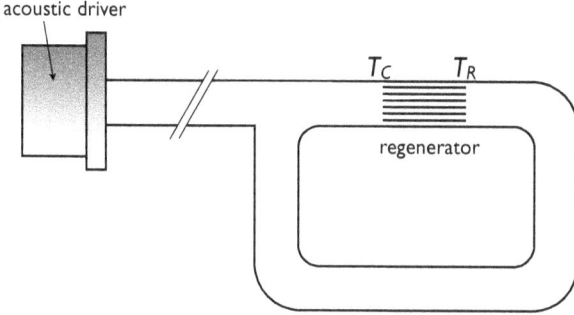

Figure 1.13: Looped tube acoustic cooler.

the working gas in the Helmholtz resonator. Thanks to the elevated mean pressure, acoustic amplitudes as high as 35 kPa was achievable. One end of the stack was cooled down by 40 K from the initial temperature. Improvements further lowered the cooling temperature down to 200 K [33].

1.4.2 *Looped tube acoustic cooler*

The looped tube acoustic cooler was developed in 1999. As shown in Fig. 1.13, the looped tube acoustic cooler is made of a looped tube and a regenerator, as the looped tube engine. Due to a better thermal contact of the working gas and the flow channels in the regenerator, a higher thermal efficiency is attainable than the resonance tube acoustic cooler [34].

A coaxial tube structure, made of inner tube and outer tube sharing the central axis, is also capable of accommodating the traveling wave acoustic field as in the looped tube [35]. One simple example is shown in Fig. 1.14, where the inner tube was eliminated for simplicity. The regenerator, a stack of screen meshes, forms a cylindrical shape whose outer diameter is smaller than the outer tube, and hence the annular region is created in the gap between the regenerator and the outer tube. In this demonstrator, a temperature difference of 20 K is attained after 30 seconds of acoustic forcing at the resonance frequency.

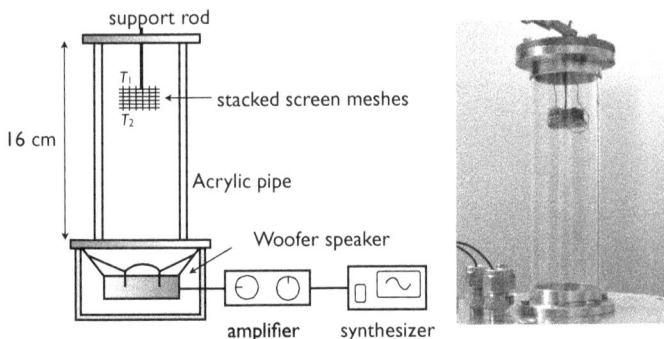

Figure 1.14: Coaxial tube type acoustic cooler.

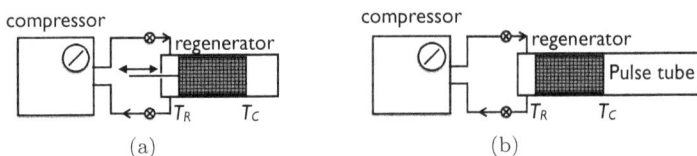

Figure 1.15: GM refrigerator (a) and basic pulse tube refrigerator (b).

1.4.3   *GM refrigerator and pulse tube refrigerator*

A GM (Gifford-McMahon) refrigerator and pulse tube refrigerator are called regenerative refrigerators, as a regenerator is an indispensable component for their operation. The working gas, driven by a compressor and valves, oscillates through the regenerator. The phasing between pressure oscillation and velocity oscillation of the gas is controlled by motion of a movable regenerator in the GM refrigerator [Fig. 1.15(a)], whereas it is adjusted by passive flow components like an orifice and a buffer tank in the pulse tube refrigerator.

Several types of flow components have been proposed. In a basic pulse tube refrigerator in Fig. 1.15(b), an empty tube called a pulse tube was connected to the regenerator. Although the cooling performance was inferior to the GM refrigerator, its simplicity attracted researchers including Wheatley. In an orifice pulse tube refrigerator in Fig. 1.16(a), a buffer tank with a sufficiently large gas volume

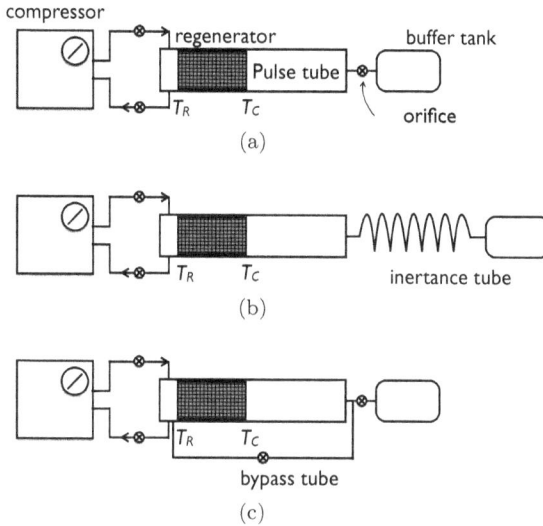

Figure 1.16: Basic types of pulse tube refrigerators: (a) orifice type, (b) inertance tube type, and (c) double inlet type.

was connected to the regenerator via an orifice, which contributed to improving performance [36]. An inertance pulse tube refrigerator [Fig. 1.16(b)] has an long tube called an inertance tube between a pulse tube and a tank [37]. A double inlet pulse tube refrigerator [Fig. 1.16(c)] has an additional bypass tube in the orifice pulse tube refrigerator [38]. The pulse tube refrigerators of the inertance and double inlet types have achieved a cooling performance comparable to the GM refrigerator, in spite of the absence of moving parts in the cold part. Through the development of pulse tube refrigerators, the importance of phasing between pressure and velocity of the working gas was accepted by engineers. Chapter 8 will explain how the passive flow components tune the acoustic field in pulse tube refrigerators.

1.4.4 *Heat driven acoustic cooler*

Acoustic coolers and pulse tube refrigerators are sourced by an acoustic driver such as a loudspeaker or a compressor, but they can also be driven by an acoustic engine. Figure 1.17(a) presents a heat driven cooler developed by Wheatley [32]. It has two stacks in

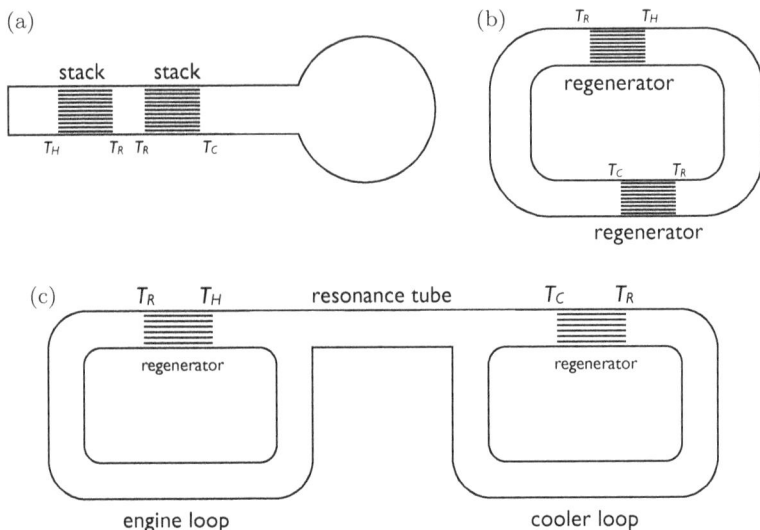

Figure 1.17: Heat driven acoustic coolers: (a) resonance tube type, (b) looped tube type and (c) double loop type.

a Helmholtz resonator, one is for producing acoustic power and the other is for making low temperatures. Therefore, it can be considered a combination of the resonance tube engine and the cooler sharing the same resonance tube. Yazaki built a heat-driven cooler [1.17(b)] made of a looped tube having a secondary regenerator in the looped tube engine [39]. It is possible to combine a looped tube engine and a looped tube cooler using a branch tube as in Fig. 1.17(c). Hasegawa succeeded in achieving a cooling temperature of $-100°C$ with a heating temperature of $300°C$ [40]. These heat-driven coolers are capable of maintaining low temperatures without using any moving parts to operate.

1.5   Dream Pipe

Kurzweg and Zhao constructed a dream pipe that consists of 31 glass tubes of 1 mm inner diameter [41]. As shown in Fig. 1.18(a), the upper end of the dream pipe is inserted into a hot water bath, while the lower end is immersed in a cold water bath equipped with

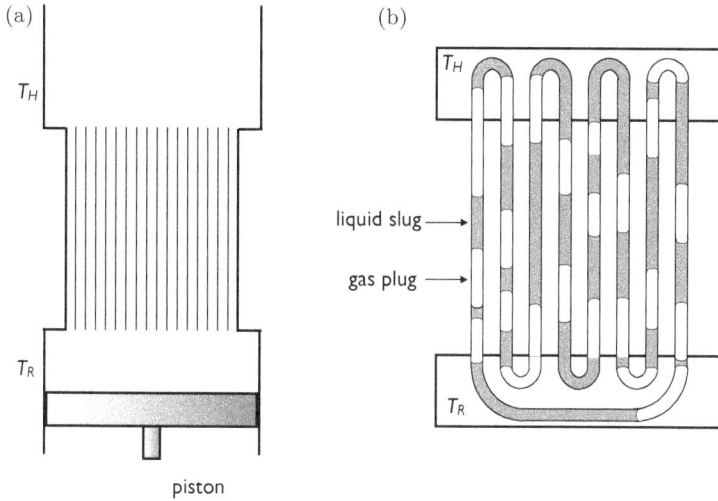

Figure 1.18: Dream pipe (a) and self-oscillating heat pipe (b).

a piston. The tubes are filled with water, and the piston vibrates the water column longitudinally. They measured temperatures of water baths after starting vibration excitation with the piston, and deduced an effective thermal diffusivity of the dream pipe. The frequency range tested was from 3 to 8 Hz, and displacement amplitude was varied in a range of 10 mm and 63 mm. They obtained a maximum effective thermal diffusivity of $25 \times 10^{-4}\,\mathrm{m^2/s}$. This value is as high as 18,000 times the thermal diffusivity of water at room temperature, and 18 times copper at room temperature. Ozawa [42] reported that the effective thermal diffusivity increased to even more than 100 times larger than that of copper using water as the working fluid.

Whereas the dream pipe is driven mechanically by a piston, it is possible to use thermally induced oscillations. In a self-oscillating heat pipe [24], a liquid slug and a gas plug occupy the internal volume of the pipe. Once an end of the pipe is heated, the oscillation of fluids starts automatically and heat is transported from hot to cold. Usually the self-oscillating heat pipe is made of a uniform tube, but the tube plays either the role of heat exchanger, regenerator, or dream pipe depending on the location.

1.6   Advantages of Thermoacoustic Devices

We have seen that thermoacoustic devices are categorized into an acoustic engine, acoustic cooler, and dream pipe. They promote generation of sound by heat, making low temperatures by sound, and heat transport by oscillating fluid, respectively. These thermoacoustic devices have several advantages over existing thermal devices as listed in the following. It would be important to find practical applications that enjoy these advantages as much as possible.

- No moving parts: the device replaces mechanical pistons with acoustic oscillations of fluid.
- Simple structure: the device is made of a few parts like heat exchanger, regenerator, and pipe.
- Less environmental problem: the device uses helium gas, air, and water as the working fluid, and the pipes are made of usual materials such as stainless steel.
- External combustion engine: the device can utilize various heat sources like solar energy, combustion heat of biomass, and waste heat from industry.
- Intrinsically high thermal efficiency: the device executes the Stirling thermodynamic cycle, which is one of reversible thermodynamic cycles.
- Wide operation temperature range: the device does not rely on the phase transition of the working fluid.

1.7   Toward Practical Application

Before closing this chapter, let's think about practical applications of thermoacoustic devices to be realized in the near future. Tijani in Energy research Centre of the Netherlands (ECN) are developing a waste heat recovery system by combining the acoustic engine and the acoustic heat pump. The acoustic engine in that system is sourced with waste heat produced in various distillation processes in a chemical plant and the heat pump driven by the engine upgrades the waste heat at lower temperature to some useful heat with higher temperature. The acoustic engine can be used in rural areas where available heat sources are limited to solar energy or biomass energy. A group at the University of Nottingham is developing a clean cooking stove that also generates electricity. The central components are an acoustic engine and a linear alternator connected to it. A heat driven cooler is particularly useful when one considers liquefying natural gases, as the heat source is also natural gas evaporated from the vessel. Los Alamos Laboratory has built a prototypical natural gas liquefier made of one acoustic engine and three pulse tube refrigerators. The achievable cooling power of 3.8 kW at 150 K would be able to liquefy 350 gallons of methane in a day.

If hydrogen becomes an important energy carrier in the future, the thermoacoustic technology to liquefy the natural gas would be applicable to liquefy hydrogen or at least to maintain the low temperature environment for sustaining liquid hydrogen. If such a technology is available, it would also help a wide application of superconductors that need low temperatures for operation in lossless electricity transport systems or storage systems. Accidents at nuclear power generation plants in 2011 exposed the potential risks. In order to maintain a cold shutdown state for all reactors to stay secured, lots of circulating cooling water is currently used and the maintenance of a filtering system is unavoidable to remove reactive substances. What if one uses a dream pipe to transport heat from the central part of the reactor while net mass transport of cooling fluid is kept at a minimum? Now the understanding of thermoacoustic devices has reached a level capable of predicting performance and efficiency quantitatively. I hope this book provides a fundamental knowledge on the energy conversion and transport mechanisms of thermoacoustic devices.

Chapter 2

Wave Propagation in a Tube

Acoustic wave propagation of a gas column confined in a tube is a fundamental problem in acoustics, and also provides a starting point for thermoacoustics. In contrast to the wave propagation in a free space, the thermal interactions between the gas and the tube wall come into play. We will see that a measure of the thermal interactions is expressed by a dimensionless parameter $\omega\tau_\alpha$, where ω denotes the angular frequency of the wave, and τ_α is a relaxation time for the gas to achieve thermal equilibrium with the wall. Depending on the value of $\omega\tau_\alpha$, the thermodynamic processes of the gas can be classified into adiabatically reversible, isothermally reversible, and irreversible ones. The associated changes in propagation constant of the acoustic plane wave will be demonstrated by experimental results.

2.1 Wave Equation and Its Solution

2.1.1 *Acoustic variables and sound waves*

Consider acoustic fluctuations of a gas column confined in a cylindrical tube, as shown in Fig. 2.1. We use a cylindrical coordinate system, where the x axis is along the central axis of the tube, and r is the radial coordinate. The internal tube radius is expressed by r_0. All the acoustic variables change with time around their temporal mean values. Pressure p, density ρ, and velocity u of the gas are

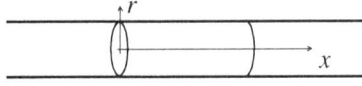

Figure 2.1: Cylindrical tube and the axial coordinate x and radial coordinate r.

expressed as follows:

$$p = p_m + p',$$ (2.1)

$$\rho = \rho_m + \rho',$$ (2.2)

$$u = u'.$$ (2.3)

The temporal mean values are expressed with a subscript m and the fluctuating components are marked with prime symbol ($'$). For example, p_m is the temporal mean pressure and p' is the acoustic pressure; u' is the acoustic particle velocity. Note that the mean velocity is assumed to be zero.

When r_0 is much shorter than the wavelength of acoustic waves, pressure fluctuation is seen as a plane wave

$$p' = p'(x, t).$$ (2.4)

In a lossless case when the gas thermal conductivity and viscosity are negligibly small, the pressure fluctuation satisfies the one-dimensional wave equation

$$\frac{\partial^2 p'}{\partial t^2} - c^2 \frac{\partial^2 p'}{\partial x^2} = 0,$$ (2.5)

where c represents the *speed of sound*. For harmonic waves with an angular frequency ω, the solution of Eq. (2.5) is given by

$$p' = P \cos(\omega t - kx),$$ (2.6)

where A represents the *amplitude* and k is the *wavenumber*,[1] and $\omega t - kx$ represents the *phase*. By inserting Eq. (2.6) to the wave equation, one can readily see the relation between c and k as

$$c = \frac{\omega}{k}.$$ (2.7)

---

[1]The wavenumber is expressed by $k = 2\pi/\lambda$, where λ is the wavelength.

Other forms of p' can satisfy the wave equation such as $p' = \exp[-(x-ct)^2]$ and $p' = \sqrt{x - ct}$. In general, the solution of Eq. (2.5) is given by using an arbitral function f as

$$p' = f(x - ct). \tag{2.8}$$

As we show in the next section, $p' = f(x-ct)$ represents the forward-traveling wave. Note that the waveform given by an arbitral function g is also a solution of the wave equation if the variable is $x + ct$:

$$p' = g(x + ct). \tag{2.9}$$

This solution means that the backward-traveling wave is going to the negative x direction with t. Therefore in general, the solution of the wave equation is given by a linear combination of forward- and backward-traveling waves

$$p' = f(x - ct) + g(x + ct). \tag{2.10}$$

2.1.2 *Wave progation*

Figure 2.2 shows how the "wave" given by the form of $p' = f(x - ct)$ in Eq. (2.8) travels in space with time. Suppose that the waveform at time $t = 0$ is $p'(x, 0) = f(x)$, as shown at the top of Fig. 2.2. At time $t = \Delta t$, Eq. (2.8) implies that $p'(x, 0)$ changes to $p'(x, \Delta t) = f(x - c\Delta t)$, which is obtained by shifting the waveform $p'(x, 0) = f(x)$ in a positive x direction by a distance of $c\Delta t$. Therefore, $p' = f(x - ct)$ represents the wave going in the positive x direction with time.

The wave propagation expressed by Eq. (2.8) can also be recognized in the following way. Let's focus on a certain point of the

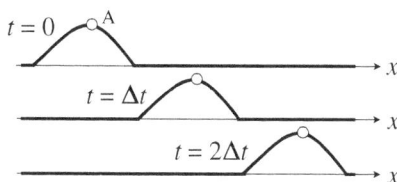

Figure 2.2: Forward-traveling wave as time passes.

waveform, for example, point A at the top of the wave in Fig. 2.2. The trajectory of the position of point A is found by letting $dp' = 0$, where

$$dp' = \frac{\partial f}{\partial t} dt + \frac{\partial f}{\partial x} dx. \qquad (2.11)$$

If we introduce a new variable $z = x - ct$, dp' is rewritten as

$$dp' = -c\frac{\partial f}{\partial z} dt + \frac{\partial f}{\partial z} dx. \qquad (2.12)$$

Hence, the relation between x and t that satisfies $dp' = 0$ is expressed as

$$\frac{dx}{dt} = c. \qquad (2.13)$$

This relation means that the wave travels with time in the direction of $+x$ at speed c. As we have noted above, the solution $p' = g(x+ct)$ represents the backward-traveling wave, which can be shown in the same way.

2.2   Speed of Sound and Acoustic Impedance

The wave equation in Eq. (2.5) can be decomposed into a pair of first order differential equations[2]

$$\rho_m \frac{\partial u'}{\partial t} + \frac{\partial p'}{\partial x} = 0, \qquad (2.16)$$

$$\frac{1}{c^2} \frac{\partial p'}{\partial t} + \rho_m \frac{\partial u'}{\partial x} = 0. \qquad (2.17)$$

---

[2]It is also possible to rewrite the wave equation as

$$\left(\frac{\partial}{\partial t} - c\frac{\partial}{\partial x} \right) \left(\frac{\partial}{\partial t} + c\frac{\partial}{\partial x} \right) p' = 0 \qquad (2.14)$$

which gives a pair of differential equations

$$\frac{\partial p'}{\partial t} + c\frac{\partial p'}{\partial x} = 0, \quad \frac{\partial p'}{\partial t} - c\frac{\partial p'}{\partial x} = 0. \qquad (2.15)$$

The solutions are $p' = f(x - ct)$ and $p' = g(x + ct)$, respectively.

Using each of these equations, let us discuss specific acoustic impedance and speed of sound, respectively.

2.2.1 Specific acoustic impedance

Equation (2.16) represents the equation of motion for a gas parcel. Figure 2.3 illustrates a short segment of a pipe, where a gas parcel moves with velocity u' by pressure difference Δp. Inserting $p' = f(x - ct)$ in Eq. (2.8) yields

$$\frac{\partial u'}{\partial t} = -\frac{1}{\rho_m}\frac{\partial p'}{\partial x} = -\frac{1}{\rho_m}\frac{\partial f}{\partial z} = \frac{1}{\rho_m c}\frac{\partial f}{\partial t}. \tag{2.18}$$

By integrating with respect to t and by setting a constant of integration to zero, we arrive at

$$u' = \frac{1}{\rho_m c}f. \tag{2.19}$$

Therefore, the velocity fluctuation u' and the pressure fluctuation p' are expressed by the same function f and a coefficient $\rho_m c$. If the pressure is a sinusoidal function of Eq. (2.6), the velocity is written as

$$u' = \frac{P}{\rho_m c}\cos(\omega t - kx). \tag{2.20}$$

The constant $\rho_m c$ is called *characteristic impedance* of the gas. The ratio of acoustic pressure p' over the acoustic particle velocity u' is called *specific acoustic impedance* $z = p'/u'$. Therefore, a forward-traveling wave going in the positive direction x, z is equal to the

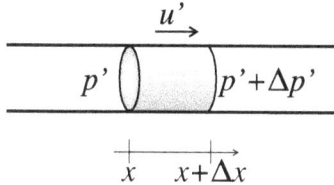

Figure 2.3: A short segment of the pipe, where a gas parcel moves with the velocity u' under a pressure difference $\Delta p'$.

characteristic impedance of the gas. Namely,

$$z = \rho_m c. \tag{2.21}$$

On the other hand, for the backward-traveling wave of $p' = g(x+ct)$, the relation

$$z = -\rho_m c \tag{2.22}$$

holds.

2.2.2 *Speed of sound*

Equation (2.17) represents the equation of continuity. Figure 2.4 depicts a short segment of the pipe, where the gas particle has different velocities at ends because of the density fluctuation. The equation of continuity is usually expressed in terms of velocity and density instead of pressure as

$$\frac{\partial \rho'}{\partial t} + \rho_m \frac{\partial u'}{\partial x} = 0. \tag{2.23}$$

From comparison of Eqs. (2.17) and (2.23), one should notice that the relation

$$c^2 \frac{\partial \rho'}{\partial t} = \frac{\partial p'}{\partial t} \tag{2.24}$$

holds. Therefore, the speed of sound is governed by the change of pressure and density as

$$c = \sqrt{\left(\frac{\partial p}{\partial \rho}\right)} \tag{2.25}$$

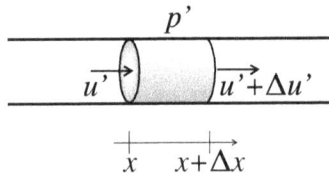

Figure 2.4: A short segment of the pipe where the gas goes in with velocity u' and goes out with $u' + \Delta u'$. The net mass entering the segment is linked with the change of density of the gas in it by the equation of continuity. The change of density is determined by pressure fluctuation p' when the thermodynamic process is known for the gas.

which indicates that the speed of sound is the quantity of thermo-dynamics rather than fluid dynamics, because the relation between pressure and density depends on the thermodynamic process that the gas experiences.

2.2.3 *Isothermal sound speed and adiabatic sound speed*

In 1697, Newton predicted the speed of sound that travels through atmospheric air, but in 1816, Laplace theoretically derived it in a different way. The difference between them originates from the assumptions for thermodynamic process. Newton's speed of sound is called isothermal speed of sound, and given by

$$c_T = \sqrt{\left(\frac{\partial p}{\partial \rho}\right)_T}. \tag{2.26}$$

The subscript T denotes the isothermal process where gas temperature is constant. His derivation of c_T was based on Boyle's law, which was the only relation between pressure and density known in that era.

Laplace's speed of sound is called adiabatic speed of sound, and given by

$$c_S = \sqrt{\left(\frac{\partial p}{\partial \rho}\right)_S}. \tag{2.27}$$

The subscript S indicates the adiabatic process where gas entropy is constant. In this case, the gas temperature temporally changes with pressure. If we assume an ideal gas, the isothermal speed of sound, c_T and the adiabatic speed of sound, c_S, are respectively given by

$$c_T = \sqrt{\frac{p_m}{\rho_m}} \tag{2.28}$$

$$c_S = \sqrt{\frac{\gamma p_m}{\rho_m}} \tag{2.29}$$

where γ represents the ratio of isobaric specific heat and isochoric specific heat.

As $\gamma = 1.4$ for air, these speeds are apparently different from each other. Namely, $c_T = 290\,\text{m/s}$ and $c_S = 340\,\text{m/s}$ at standard reference atmospheric condition. Through comparison with measurements, it turns out that the adiabatic speed of sound c_S better described the actual propagation speed in Laplace's era [43]. Since then, the adiabatic process has been well accepted as a reasonable assumption for the problem of acoustic wave propagation, although it is not capable of explaining any thermoacoustic phenomena caused by the interaction between sound and heat.

2.3 Propagation Constant for Acoustic Wave in a Tube

2.3.1 *Temperature fluctuation of sound waves*

The change of gas entropy S is expressed by the pressure change and temperature change as

$$dS = \left(\frac{\partial S}{\partial p}\right)_T dp + \left(\frac{\partial S}{\partial T}\right)_p dT. \tag{2.30}$$

The partial derivatives are rewritten by using thermodynamic relations as

$$\left(\frac{\partial S}{\partial T}\right)_p = \frac{C_p}{T_m}, \tag{2.31}$$

$$\left(\frac{\partial S}{\partial p}\right)_T = -\left(\frac{\partial V}{\partial T}\right)_p = -\frac{\beta}{\rho_m}, \tag{2.32}$$

where $\beta = -\frac{1}{V}\left(\frac{\partial V}{\partial T}\right)_p$ denotes the thermal expansion coefficient. In the adiabatic process with $dS = 0$, one obtains

$$dT = \frac{\beta T_m}{\rho_m C_p} dp. \tag{2.33}$$

This equation means that the gas temperature changes in proportion to the pressure. Therefore, the temperature goes up when the pressure increases, and vice versa. In a free space, the thermal conduction associated with the pressure-induced temperature gradient is readily small because the spatial pressure change takes place only in the scale of wavelength, except for the vicinity of shock fronts in high-amplitude sound waves. Therefore, the gas temperature changes in time but rather uniformly in space. In this case, the adiabatic approximation is a reasonable assumption, because the gas parcel does not have any counterpart to exchange heat with.

In the presence of solid walls, however, the thermodynamic processes are influenced by it, particularly in the vicinity of the wall. The solid wall normally has a sufficiently high heat capacity so that it can act as a heat bath maintaining constant temperature. Therefore, the gas parcel in contact with the wall always keeps a constant temperature against the pressure change. As a result, the gas parcel experiences isothermal process at the wall surface, whereas it experiences adiabatic process far from the wall surface. Also, for the gas in the intermediate region between the regions of adiabatic process and isothermal process, the irreversible thermodynamic process of thermal conduction takes place owing to the temperature distribution in a direction normal to the wall.

2.3.2 *Measure of thermal contact of gas with wall*

Let's consider here the thermodynamic process of the gas near the wall through an intuitive discussion based on dimensional analysis. (In Chapter 5 we will analyze this problem in more detail using the solution of basic equations of hydrodynamics.) The angular frequency ω of the acoustic wave has the unit of $(1/s)$ and the thermal diffusivity α of the gas has the unit of (m^2/s). By using ω and α we obtain the characteristic length

$$\delta_\alpha = \sqrt{\frac{2\alpha}{\omega}} \qquad (2.34)$$

which is known as the *thermal boundary layer thickness*. The thermal boundary layer thickness δ_α gives a measure of distance from the wall within which a gas parcel oscillates under the influence of thermal contact with the solid wall. Suppose that the gas parcel oscillates above a semi-infinite solid plate shown in Fig. 2.5. For the gas parcel

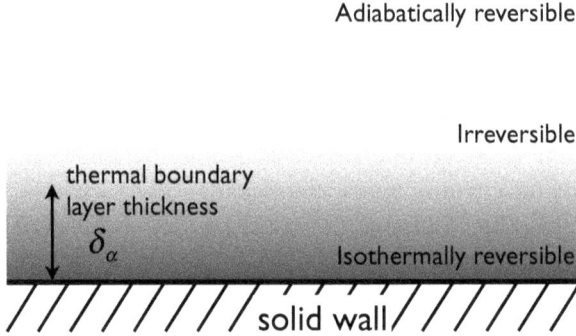

Figure 2.5: Semi-infinite solid plate and gas. Thermodynamic process of the gas gradually changes from the isothermal one to the adiabatic one as the distance from the plate goes beyond δ_α.

oscillating along the plate keeping a distance much less than δ_α, we can assume the isothermal thermodynamic process. On the other hand, the gas parcel adiabatically oscillates when the distance from the plate is much more than δ_α. For the intermediate distance about δ_α from the wall surface, the gas parcel experiences irreversible thermodynamic process since it is exposed to the transverse temperature gradient all the time. In the same way, for the gas confined in a tube with radius r_0, the ratio r_0/δ_α can be seen as a parameter governing the nature of thermodynamic process of the gas.

If we use the thermal diffusivity α and the tube radius r_0, we can introduce a characteristic time

$$\tau_\alpha = \frac{r_0^2}{2\alpha} \qquad (2.35)$$

which is known as a transversal *thermal relaxation time* of the gas. It gives a measure of how much time is necessary for the gas to achieve a thermal equilibrium in the cross section of the tube when the gas is thermally perturbed. When the condition $\omega\tau_\alpha \ll 1$ is achieved, the thermal equilibrium is always maintained and the gas temperature remains the same as the tube wall temperature. On the other hand, when $\omega\tau_\alpha \gg 1$ is achieved, the gas temperature adiabatically oscillates with pressure obeying Eq. (2.33). For the intermediate values of $\omega\tau_\alpha$, the thermodynamic process becomes irreversible.

The non-dimensional parameters $\omega\tau_\alpha$ and r_0/δ_α are linked to each other through the relation

$$\omega\tau_\alpha = \left(\frac{r_0}{\delta_\alpha}\right)^2.$$ (2.36)

It is worthwhile mentioning that $\omega\tau_\alpha$ and r_0/δ_α are mathematically equivalent, but their physical meanings are different; r_0/δ_α refers to spatial behavior, while $\omega\tau_\alpha$ addresses temporal behavior. Figure 2.6 illustrates a schematic view of the cross-section of the pipes. It should be noted that when $\omega\tau_\alpha \ll 1$, the temperature is always uniform over the cross-section of the tube, and therefore, there exists no thermal boundary layer. In this particular case, one cannot talk about the thermal boundary layer thickness.

For air at ambient pressure and 20°C, the thermal diffusivity is $\alpha = 21.1 \times 10^{-6}\,\mathrm{m^2/s}$. When $r_0 = 1\,\mathrm{mm}$, τ_α becomes 0.023 s. Therefore, the condition $\omega\tau_\alpha = 1$ is achieved when the frequency is 6.7 Hz. The thermodynamic processes of the gas go to the isothermal one if the frequency is reduced, and the adiabatic one if the frequency is raised.

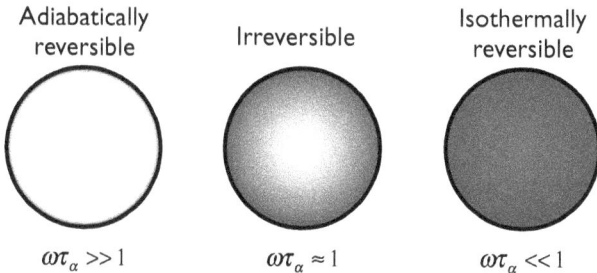

Figure 2.6: Schematic cross-sectional view of a cylindrical tube. Thermodynamic processes of the gas is represented by colors. The gas is strongly influenced by the wall in the gray regions, whereas it is less influenced in the white regions.

2.3.3 *Plane pressure waves in tube*

Because the gas thermodynamic process is influenced by the presence of the wall, one can easily imagine that the speed of sound in a gas-filled tube differs from the adiabatic speed of sound c_S or the isothermal speed of sound c_T. In 1975, Tijdeman theoretically analyzed the acoustic wave propagation in a gas-filled cylinder [44]. His results showed that the acoustic pressure wave of an angular frequency ω was expressed by

$$p' = \text{Re}[P \exp{(i\omega t - k_S \Gamma x)}] \tag{2.37}$$

where k_S is the adiabatic wavenumber $(k_S = \omega/c_S)$ and Γ is a complex quantity given by

$$\Gamma = \sqrt{-\frac{1 + (\gamma - 1)\chi_\alpha}{1 - \chi_\nu}}. \tag{2.38}$$

Here, χ_i $(i = \alpha, \nu)$ is a complex quantity specified by using the thermal diffusivity α or the kinematic viscosity ν. In this section, we point out the impact of thermal interaction on sound propagation. Derivations of Γ and χ_i $(i = \alpha, \nu)$ are explained later on the basis of the basic equations of fluid dynamics.

By comparison with an alternative expression $p' = \text{Re}[P \exp i(\omega t - kx)]$, Γ is rewritten by using the wavenumber k as

$$\Gamma = i\frac{k}{k_S}. \tag{2.39}$$

This relation indicates that the wavenumber becomes a complex quantity in the tube, while it is a real quantity in a free space. If we decompose k into the real part $\text{Re}\,k$ and the imaginary part $\text{Im}\,k$ as $k = \text{Re}\,k + i\text{Im}\,k$, the pressure wave is expressed as

$$p' = P \exp{[(\text{Im}\,k)x]}\,\text{Re}[\exp i\{\omega t - (\text{Re}\,k)x\}]. \tag{2.40}$$

If we use an attenuation constant ς and a phase velocity v, the plane wave is expressed by

$$p' = P \exp{(-\varsigma x)}\,\text{Re}[\exp i\{\omega(t - x/v)\}]. \tag{2.41}$$

Therefore, Γ and k are linked with ς and υ as

$$\varsigma = -\operatorname{Im} k = k_S \operatorname{Re} \Gamma \qquad (2.42)$$

$$\upsilon = \frac{\omega}{\operatorname{Re} k} = \frac{c_S}{\operatorname{Im} \Gamma} \qquad (2.43)$$

2.3.4 Phase velocity and attenuation constant in a tube

Figure 2.7 illustrates the phase velocity υ [Eq. (2.42)] and the attenuation constant ς [Eq. (2.43)], where υ is normalized with respect to the adiabatic speed of sound, and ς is divided by adiabatic wavenumber k_S.

Let's focus on asymptotic behavior of the phase velocity υ and the attenuation constant ς when $\omega\tau_\alpha$ becomes infinitely large or small. The asymptotic curve with $\omega\tau_\alpha \gg 1$ was derived by Kirchhoff based on a "wide tube" approximation, where the tube radius is assumed to be much larger than the boundary layer thickness. The curve shown

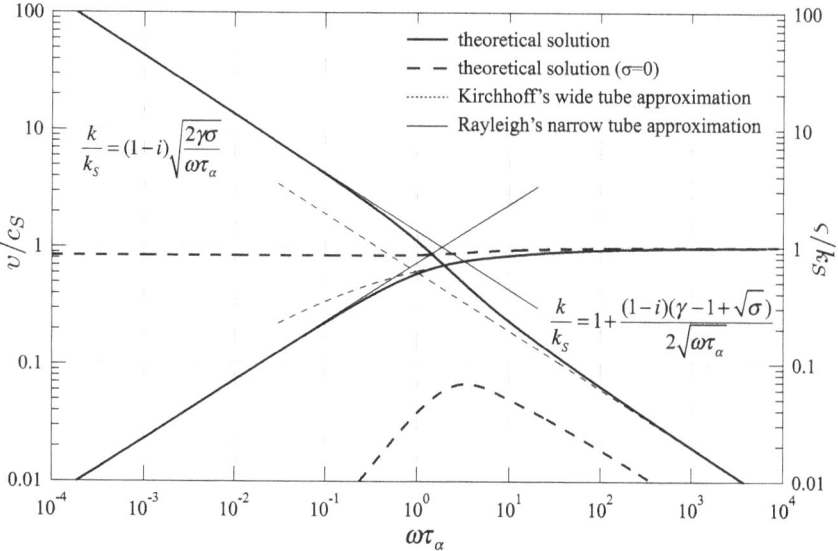

Figure 2.7: Phase velocity υ and attenuation constant ς as a function of $\omega\tau_\alpha$.

by thin dashed lines in Fig. 2.7 is given as

$$\frac{k}{k_S} = 1 + \frac{(1-i)(\gamma - 1 + \sqrt{\sigma})}{2\sqrt{\omega \tau_\alpha}}, \qquad (2.44)$$

where γ is the specific heat ratio of the gas and σ is the Prandtl number.

In the region with $\omega \tau_\alpha \gg 1$, it is seen from Eq. (2.44) that $(\mathrm{Re}\,k)/k_S \approx 1$, and $(\mathrm{Im}\,k)/k_S \approx (\gamma - 1 + \sqrt{\sigma})/(2\sqrt{\omega \tau_\alpha})$ hence the phase velocity $v = \omega/\mathrm{Re}\,k$ approaches the adiabatic speed of sound c_S and the attenuation constant ς becomes infinitely small. Thus, in the limit of the wide tube, the acoustic wave in a tube is seen as the adiabatic acoustic wave in a free space. The Kirchhoff's solution of Eq. (2.44) practically reproduced the exact solution given in Eq. (2.38) down to $\omega \tau_\alpha \approx 2$ for the phase velocity but $\omega \tau_\alpha \approx 10$ for the attenuation constant. Lord Rayleigh introduced the Kirchhoff theory in his treatise "The Theory of Sound" [1], and he obtained the asymptotic curve when $\omega \tau_\alpha \ll 1$ based on a narrow tube approximation that becomes valid for the tube with $r_0 \ll \delta_\alpha$. The approximate solution is given as

$$\frac{k}{k_S} = (1-i)\sqrt{\frac{2\gamma\sigma}{\omega \tau_\alpha}}. \qquad (2.45)$$

As shown by thin curves in Fig. 2.7, Rayleigh's solution gives a good approximation in the region with $\omega \tau_\alpha < 0.4$. The phase velocity decreases below c_T with decreasing $\omega \tau_\alpha$, although the isothermal process should be realized when $\omega \tau_\alpha \ll 1$. The speed of sound decreases down to 1% of c_S ($\sim$12 kilometers per hour; can you run faster than sound?) because of the unavoidable viscous effect of the gas. Indeed, the attenuation constant grows considerably in the low $\omega \tau_\alpha$ region. If we set $\nu = 0$ in Eq. (2.38), the phase velocity v and the attenuation constant ς are given as thick dashed lines. The attenuation constant ς starts to decrease as the isothermal process is achieved with lower values of $\omega \tau_\alpha$, and the phase velocity v reaches the isothermal sound speed c_T that Newton derived.

The solutions of Kirchhoff and Rayleigh fit the exact solution in wide areas of $\omega \tau_\alpha$, but the asymptotic curves both fail in the region near $\omega\tau_\alpha \approx 1$. In this region, the gas experiences the irreversible thermodynamic process where neither isothermal condition nor adiabatic condition can be applied. As a result, whereas the viscous damping is moderate, the heat exchange takes place between the gas and the tube wall, which makes it possible for various thermoacoustic phenomena to occur in the intermediate region of $\omega\tau_\alpha$.

2.4 Appendix

2.4.1 *Minimum audible sound*

A minimum audible sound of a pure tone that a human can hear is reported as 2×10^{-5} Pa at 1 kHz. Since the specific acoustic impedance $\rho_m c$ is 415 Pa s/m for air at atmospheric pressure and at 15°C, the acoustic particle velocity amplitude can be estimated as

$$|u'| = \frac{2 \times 10^{-5}}{415} \approx 4.8 \times 10^{-8} \, \text{m/s}. \tag{2.46}$$

The displacement of the gas particle is given by integration of the velocity with respect to time. Since we are dealing with harmonic oscillations, the gas displacement amplitude is simply given by $|u'|/\omega$. So it turns out that

$$|\xi'| = \frac{4.8 \times 10^{-8}}{2\pi \times 10^3} \approx 0.76 \times 10^{-11} = 0.0076 \, \text{nm}. \tag{2.47}$$

The obtained displacement is much shorter than the radius of Hydrogen atoms (Bohr radius, 0.053 nm). Human's ear is surprisingly sensitive.

2.4.2 *Experiments of wave propagation in a tube*

The propagation constant given by Eq. (2.38) was recently verified by experiments [45]. The schematic of experimental setup is shown in Fig. 2.8. The tube is so long that the reflection at the right end is negligibly small and only a forward-traveling wave is created by a loudspeaker at the left end. The governing parameter $\omega\tau_\alpha$ of Γ is changed by using tubes with three different internal radius r_0 of 0.3, 1.0, and 2.0 mm, and also by varying the kinematic viscosity ν of

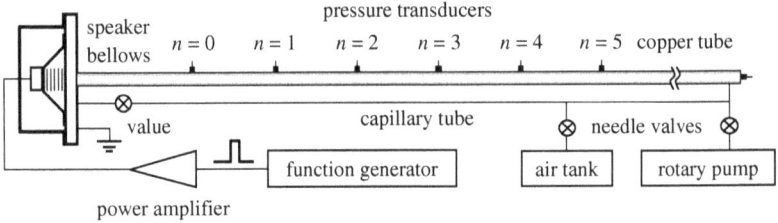

Figure 2.8: Experimental setup for measuring the sound wave propagation in a tube.

the gas through the mean pressure dependence. The values of mean pressure p_m are 2.5, 5.0, 10.2, or 40.6 kPa.

A short acoustic pulse of about 10 Pa was generated from the loudspeaker and its spatiotemporal evolution was measured by a series of pressure transducers located at $x_1, x_2, \ldots, x_5$ as it traveled down the tube. The measured pressure can be seen as a combination of acoustic fluctuations of various frequencies, each of which can be written as the forward-traveling plane wave of the form in Eq. (2.37). Namely, the Fourier component $P_i(\omega)$ of pressure p_i measured at positions x_i ($i = 1, 2, \ldots, 5$) are considered to have a factor $\exp(-k_0 \Gamma x_i)$. Therefore, the transfer function $G(\omega) = P_2(\omega)/P_1(\omega)$ is expressed as

$$G(\omega) = \exp[-k_S d \operatorname{Re}\Gamma] \exp[i(-k_S d \operatorname{Im}\Gamma)], \qquad (2.48)$$

where d represents the distance between x_2 and x_1. From the magnitude of $G = \exp[-k_S d \operatorname{Re}\Gamma]$ and the phase shift $\Phi = -k_S d \operatorname{Im}\Gamma$, we see that the attenuation constant ς and the phase speed v are respectively determined as

$$\varsigma = -\frac{\log G(\omega)}{d} \qquad (2.49)$$

$$v = \omega d/\Phi \qquad (2.50)$$

The obtained results are shown in Fig. 2.9, where the value of $\omega \tau_\alpha$ extends over seven orders from 10^{-4} to 10^2. We see that ς and v go onto universal curves having a single parameter $\omega \tau_\alpha$ even though r_0 and ν are different. This result serves as experimental evidence

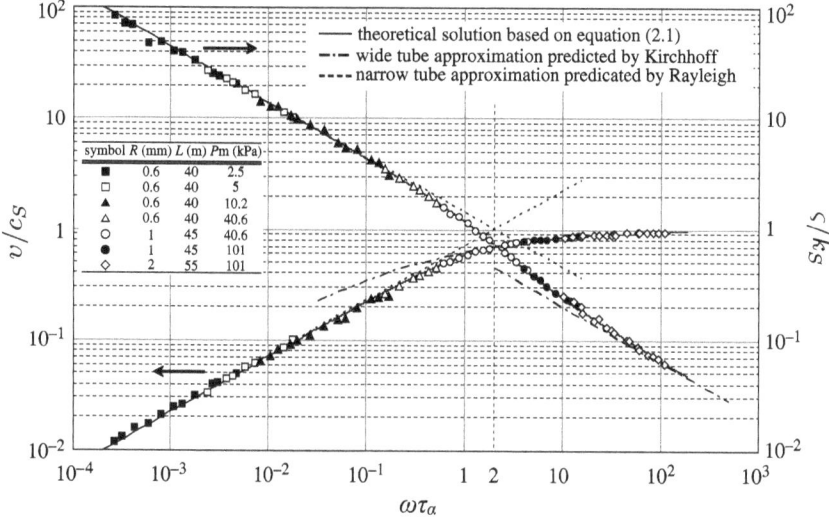

Figure 2.9: Experimental phase velocity v and attenuation constant ς as a function of $\omega\tau_\alpha$.

showing that $\omega\tau_\alpha$ governs the acoustic wave propagation, as indicated by Tijdemann. It is surprising that a fundamental problem of acoustic wave propagation was finally verified by this experiment in the 21st century.

2.4.3 *Dimensional analysis* $-\omega\tau_\nu$ *and* $Re-$

We introduced a dimensionless parameter $\omega\tau_\alpha$ for describing the acoustic wave in a tube. Another parameter given by

$$\omega\tau_\nu = \omega\frac{r_0^2}{2\nu} \qquad (2.51)$$

also serves as a characteristic dimensionless parameter. It is easy to see from their definitions that $\omega\tau_\alpha$ and $\omega\tau_\nu$ are converted to each other through the Prandtl number σ of the gas as

$$\omega\tau_\nu = \frac{\omega\tau_\alpha}{\sigma}. \qquad (2.52)$$

As σ is of the order of unity for usual gases, $\omega\tau_\alpha$ and $\omega\tau_\nu$ are close to each other. Let's find out how many dimensionless parameters are necessary for identifying the acoustic waves in a tube by applying dimensional analysis.

Consider the acoustic waves of an angular frequency ω and an acoustic particle velocity U of the gas with a kinematic viscosity ν confined in a tube of diameter d. If we adopt fundamental units of length [L] and time [T], the respective units are given as U [L/T], d [L], ν [L^2/T], and ω [1/T]. Any physical variable $U^A d^B \nu^C \omega^D$ determined by a product of these representative quantities should have a unit of

$$\left(\frac{L}{T}\right)^A L^B \left(\frac{L^2}{T}\right)^C \left(\frac{1}{T}\right)^D = \frac{L^{A+B+2C}}{T^{A+C+D}}. \tag{2.53}$$

So it becomes non-dimensional if the following relations are satisfied

$$A + B + 2C = 0 \tag{2.54}$$
$$A + C + D = 0. \tag{2.55}$$

The number of representative physical quantities is four while the number of equations is two. Owing to two degrees of freedom, there should be two individual non-dimensional parameters for describing the system. If we look for a non-dimensional parameter that is free from ω, we set D = 0 in the coupled equations above and we have

$$A + B + 2C = 0 \tag{2.56}$$
$$A + C = 0 \tag{2.57}$$

By setting A = 1 yields B = 1 and C = −1. The combination of these gives the Reynolds number

$$\mathrm{Re} = \frac{Ud}{\nu}. \tag{2.58}$$

On the other hand, the non-dimensional parameter free from ν is obtained in the same way as

$$\mathrm{St} = \frac{\omega d}{U}, \tag{2.59}$$

which is called Strauhal number. The product of Re and St also gives the non-dimensional parameter: $\text{ReSt} = \omega d^2/\nu$ which yields the dimensionless parameter free from U that is eight times larger than

$$\omega\tau_\nu = \frac{\omega r_0^2}{2\nu}. \tag{2.60}$$

In this way, $\omega\tau_\nu$ is given by Re and St, meaning that the independent numbers of dimensionless parameters are two. Any two of $\omega\tau_\nu$, Re, and St can be selected depending on one's purpose. For acoustic waves with infinitely small amplitudes, the dimensionless parameter $\omega\tau_\nu$ (or $\omega\tau_\alpha$) plays a fundamental role. Also for the acoustic waves with very large amplitudes, Re is used to characterize acoustic turbulence. Ohmi *et al.* [46–49] mapped out the flow patterns in acoustic waves in Fig. 2.10. The map indicates that oscillating fluid flow becomes turbulent when Re is raised to a certain level, although the threshold changes depending on $\omega\tau_\nu$.

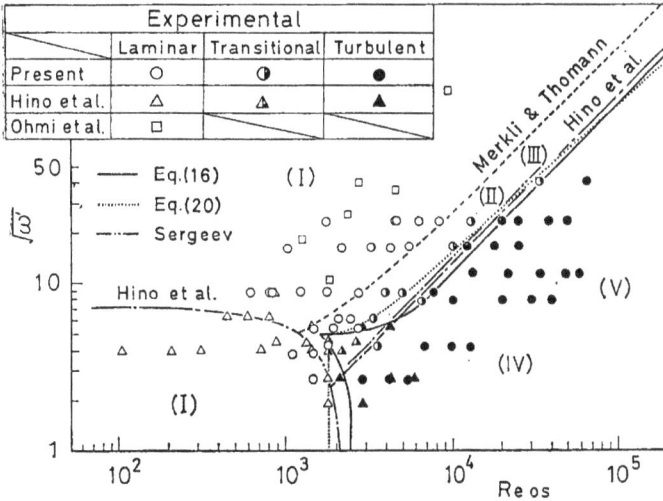

Figure 2.10: Map of oscillatory flow states obtained by Ohmi *et al.* [46–49]. Symbols $\sqrt{\omega'}$ and Re_{os} are $\sqrt{\omega'} = \sqrt{2\omega\tau_\nu}$ and $\text{Re}_{os} = |u_{1r}|d/\nu$, where $|u_{1r}|$ is the amplitude of cross-sectional mean velocity and d is the tube diameter.

2.5 Problems

1 Show that the isothermal speed of sound and the adiabatic one are expressed by $c_T = \sqrt{\dfrac{p}{\rho}}$ and $c_S = \sqrt{\dfrac{\gamma p}{\rho}}$ for an ideal gas, respectively.

The isothermal speed of sound is given by $c_T = \sqrt{\left(\dfrac{\partial p}{\partial \rho}\right)_T}$. From the state equation $p = \rho RT$, we have $\left(\dfrac{\partial p}{\partial \rho}\right)_T = RT = \dfrac{p}{\rho}$. Therefore, $c_T = \sqrt{\dfrac{p}{\rho}}$.

The adiabatic speed of sound is given by $c_S = \sqrt{\left(\dfrac{\partial p}{\partial \rho}\right)_S}$. From the first law of thermodynamics, we have $\delta q = du + p\,dv$. By letting $\delta q = 0$ and using $du = C_V dT$, we obtain $C_V dT + p\,dv = 0$, which can be written as $\dfrac{C_V}{R} d(pv) + p\,dv = 0$ by the state equation. Therefore, we have $\gamma \dfrac{dv}{v} + \dfrac{dp}{p} = 0$ and hence $pv^\gamma = const.$ by integration. From this relation, we arrive at $c_S = \sqrt{\dfrac{\gamma p}{\rho}}$ after some calculations.

2 Show that $\left(\dfrac{\partial T}{\partial p}\right)_S = \dfrac{\gamma - 1}{\gamma}\dfrac{T}{p}$ for an ideal gas.

The first law of thermodynamics is written as $dh = TdS + vdp$, where h denotes the specific enthalpy. For the ideal gas, we have $dh = C_p dT$. Therefore, by letting $dS = 0$, we obtain $\left(\dfrac{\partial T}{\partial p}\right)_S = \dfrac{v}{C_p} = \dfrac{RT}{pC_p}$, where the state equation was used in the last formation. Inserting Mayer's formula $R = C_p - C_v$ yields $\left(\dfrac{\partial T}{\partial p}\right)_S = \dfrac{C_p - C_v}{C_p}\dfrac{T}{p} = \dfrac{\gamma - 1}{\gamma}\dfrac{T}{p}$.

Chapter 3

Quality Factor of Acoustic Resonance Tube

When a gas column with finite length is driven by an external periodic force, acoustic resonance phenomenon takes place at a certain frequency. The quality factor is a dimensionless quantity that characterizes the resonance from an energetic point of view; the higher quality factor indicates the lower energy dissipation rate. In this chapter, we begin with the quality factor of a mass-spring system and then of a gas column confined in a uniform temperature tube. The reason for the appearance of thermoacoustic oscillations will be explained from the quality factor when the resonance tube has a non-zero temperature gradient.

3.1 Quality Factor

3.1.1 *Role of quality factor*

Suppose that an oscillator or resonator is driven by an external periodic force. When the driving angular frequency is equal to the natural angular frequency ω_0 of the driven system, the quality factor Q is expressed by

$$Q = \omega_0 \frac{E}{W}, \tag{3.1}$$

where E is the time-averaged energy stored in the system, and W denotes the time-averaged power supplied to the system. If the time averaged energy dissipation per unit time is denoted by $\dot{E} = dE/dt (< 0)$,

$$\dot{E} = -W, \qquad (3.2)$$

must hold in the steady state. In a lossless system with $\dot{E} = 0$, the power supply is not required ($W = 0$), and therefore, the quality factor Q becomes infinitely large. In a real system, however, the energy dissipation is unavoidable. Through Eqs. (3.1) and (3.2), we find

$$Q = -\omega_0 \frac{E}{\dot{E}}, \qquad (3.3)$$

which means that Q increases as the rate of energy dissipation becomes small. In this way, it can be said that Q represents the smallness of energy dissipation in oscillatory systems.

3.1.2 *Free oscillations and Q*

The quality factor Q is used for free oscillations, as well as for forced oscillations. After an initial disturbance, the underdamped system starts to reach the equilibrium state by reducing the oscillation amplitude while keeping the angular frequency close to ω_0. During this process, the energy E decreases obeying

$$\frac{dE}{dt} = \dot{E}. \qquad (3.4)$$

By using Eq. (3.3), we have the relation

$$\frac{dE}{dt} = -\frac{\omega_0}{Q} E, \qquad (3.5)$$

which can be solved with respect to E as

$$E = E_i \exp\left(-\frac{t}{\tau_E}\right) \qquad (3.6)$$

where the relaxation time τ_E is given by

$$\tau_E = \frac{Q}{\omega_0}, \qquad (3.7)$$

and E_i denotes a certain initial value. When we express the oscillation period by $T_{\text{osc}}(= 2\pi/\omega_0)$, Q is rewritten as

$$Q = \omega_0 \tau_E = 2\pi \frac{\tau_E}{T_{\text{osc}}}. \tag{3.8}$$

The factor τ_E/T_{osc} indicates that Q gives a measure of how many oscillations would be observed before it ceases.

3.1.3 *Q of damped oscillation system*

Consider a quality factor Q of an underdamped oscillation system as shown in Fig. 3.1. The system consists of a trailer with mass M, spring with spring constant K, and a dashpot with viscous damping coefficient R. We express by ξ' the displacement of the trailer from the equilibrium position. The equation of motion of the system reads

$$M\frac{d^2\xi'}{dt^2} + R\frac{d\xi'}{dt} + K\xi' = 0. \tag{3.9}$$

Before solving the equation of motion, we look at how the energy E

$$E = \frac{1}{2}M\left(\frac{d\xi'}{dt}\right)^2 + \frac{1}{2}K\xi'^2 \tag{3.10}$$

of the system changes with time. To do this, we multiply $d\xi'/dt$ with both sides of the equation of motion, and we have

$$M\frac{d\xi'}{dt}\frac{d^2\xi'}{dt^2} + K\xi'\frac{d\xi'}{dt} = -R\left(\frac{d\xi'}{dt}\right)^2.$$

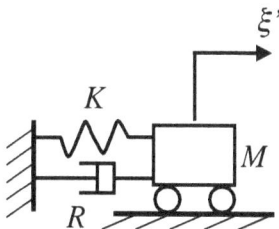

Figure 3.1: Damped oscillation system.

One can notice that the left-hand side equals to the time derivative dE/dt of the energy E in Eq. (3.10). Therefore, the following relation is obtained from the equation of motion

$$\frac{dE}{dt} = -R\left(\frac{d\xi'}{dt}\right)^2 < 0,$$

which means that the energy E must decrease with time because of the energy dissipation due to the dashpot.

Now let's solve the equation of motion directly. For simplicity, we introduce the symbols $\tau_A = 2M/R$ and $\omega_0^2 = K/M$ to rewrite the equation of motion in the following way.

$$\frac{d^2\xi'}{dt^2} + \frac{2}{\tau_A}\frac{d\xi'}{dt} + \omega_0^2\xi' = 0. \tag{3.11}$$

It should be noted that $\omega_0 = \sqrt{K/M}$ represents the natural angular frequency of the undamped system without the dashpot.

Inserting $\xi' = \exp \lambda t$ into the equation above yields a characteristic equation. From the solution, one finds

$$\xi'(t) = Ce^{-t/\tau_A}\cos\omega t \tag{3.12}$$

when R is relatively small as to satisfy $R/(2M) = 1/\tau_A < \omega_0$. Here, ω is given by $\omega = \sqrt{\omega_0^2 - (1/\tau_A)^2}$, and initial conditions are assumed as $\xi'(0) = C$, $d\xi'(0)/dt = 0$, where C is a constant. As shown in Fig. 3.2, $\xi'(t)$ represents the underdamped oscillations whose instantaneous amplitude decays exponentially with time; τ_A represents the relaxation time of the amplitude.

The energy E of the system is expressed as a function of time if we use the solution for Eq. (3.12) into Eq. (3.10);

$$E(t) = \frac{1}{2}M\left[Ce^{-t/\tau_A}\left(\frac{1}{\tau_A}\cos\omega t + \omega\sin\omega t\right)\right]^2$$

$$+ \frac{1}{2}K\left[Ce^{-t/\tau_A}\cos\omega t\right]^2.$$

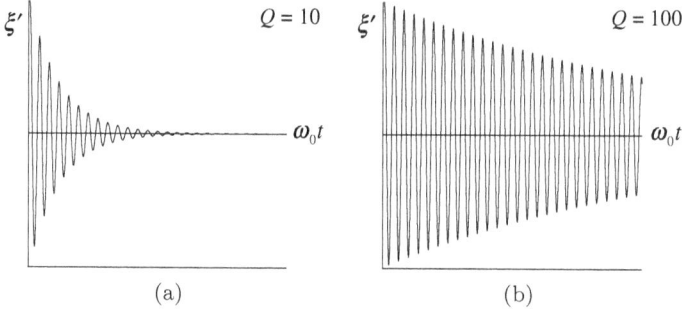

Figure 3.2: Temporal change of damped oscillation ξ'. (a) $Q = 10$ and (b) $Q = 100$.

Because $\cos^2 \omega t = (1 + \cos 2\omega t)/2$, and $\sin \omega t \cos \omega t = (\sin 2\omega t)/2$, we see that E fluctuates with angular frequency 2ω as it damps out. The time average of dE/dt over a period is given by

$$\frac{dE}{dt} = \frac{E(t + T_{\text{osc}}) - E(t)}{T_{\text{osc}}}, \tag{3.13}$$

where

$$E(t + T_{\text{osc}}) = E(t)e^{-2T_{\text{osc}}/\tau_A}. \tag{3.14}$$

Therefore,

$$\frac{dE}{dt} = \frac{E(t)}{T_{\text{osc}}}\left[e^{-2T_{\text{osc}}/\tau_A} - 1\right] \approx \frac{E(t)}{T_{\text{osc}}}\left(1 - \frac{2T_{\text{osc}}}{\tau_A} + \cdots - 1\right)$$

$$= -\frac{2}{\tau_A}E(t).$$

Finally, we obtain the relation

$$\frac{dE}{dt} = -\frac{2}{\tau_A}E(t). \tag{3.15}$$

If this equation is compared with Eq. (3.5), we see $Q = \omega_0 \tau_A/2$ or $\tau_A = 2Q/\omega_0$. Therefore, the relaxation time τ_E of energy is linked with the relaxation time τ_A of amplitude as $\tau_A = \tau_E/2$.

As $\tau_A = 2M/R$ and $Q = \omega_0 \tau_E$, Q is expressed as

$$Q = \frac{\sqrt{MK}}{R} \tag{3.16}$$

for the mechanical system shown in Fig. 3.1. It should be noted here again that Q decreases as R increases and vice versa.

3.2 Complex Representation of Oscillations

3.2.1 *Phasor diagram*

In order to prepare for discussion of the quality factor in a periodically steady system, we introduce a complex representation of oscillations, as it offers a useful mathematical tool.

Consider a physical quantity X' that oscillates with time at a single angular frequency ω. By denoting its amplitude A $(A > 0)$ and its initial phase θ, it is written as

$$X' = A\cos(\omega t + \theta). \tag{3.17}$$

Euler's formula describes the relationship between the trigonometric functions with the exponential function:

$$e^{i\phi} = \cos\phi + i\sin\phi, \tag{3.18}$$

where i denotes the imaginary unit $(i^2 = -1)$. Equation (3.18) is rewritten as

$$\cos\phi = \text{Re}[e^{i\phi}]$$
$$\sin\phi = \text{Im}[e^{i\phi}]$$

where Re[ ] and Im[ ] mean the real part and the imaginary part of the internal complex number, respectively. Therefore, we are able to express X' as

$$X' = \text{Re}[X_1 e^{i\omega t}], \tag{3.19}$$

where

$$X_1 = Ae^{i\theta} \tag{3.20}$$

is called the *complex amplitude* of X'; X_1 includes both the amplitude A and the initial phase θ.

Use of the complex amplitude allows us to graphically express the oscillations in a complex plane. Complex number $X_1 e^{i\omega t}$ is represented by a point that rotates on a circle of radius A with time-dependent phase $\omega t + \theta$. A vector-like quantity directed from the origin to that point is called a phasor. As shown in Fig. 3.3(a), the phasor is represented as a rotating vector whose tip goes around the origin with time. The projection of the phasor onto the real axis gives the oscillations X'. In other words, the phasor representation converts the one-dimensional oscillatory motion into the two-dimensional rotating motion.

It is often the case that X_1 is plotted as a static vector by introducing the coordinates that rotate with angular frequency ω, as shown in Fig. 3.3(b). Such treatment is allowed because all the oscillating quantities share the same angular frequency as long as we consider linear oscillation systems.

The graphical representation of oscillations has two advantages. Firstly, it is easy to obtain an intuitive understanding of the phasing between oscillations. For example, suppose that an oscillating quantity $Y' = \text{Re}[Y_1 e^{i\omega t}]$ is related with X' through

$$Y_1 = \Gamma X_1, \qquad (3.21)$$

where $\Gamma = |\gamma| e^{i\Theta}$ is a complex constant. The amplitude ratio $|\gamma|$ of Y' to X' and the phase lead Θ of Y' relative to X' are obtained from

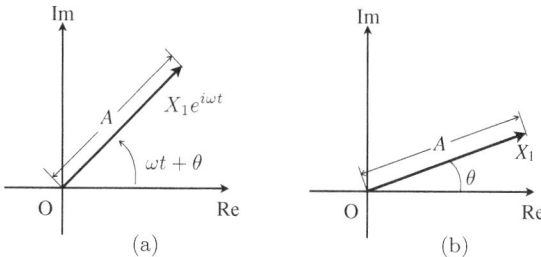

Figure 3.3: Phasor diagram.

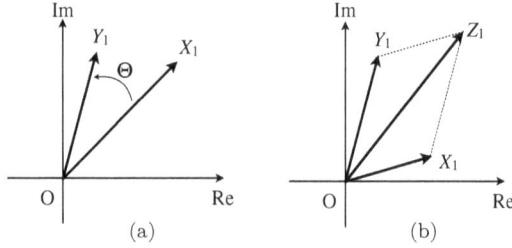

Figure 3.4: Phasor diagram: (a) $Y_1 = \Gamma X_1$ and (b) $Z_1 = X_1 + Y_1$.

calculations such as the following

$$
\begin{aligned}
Y' &= \mathrm{Re}[\Gamma X_1 e^{i\omega t}] \\
&= \mathrm{Re}[|\gamma||A|e^{i\Theta}e^{i\theta}e^{i\omega t}] \\
&= |\gamma|A\cos(\omega t + \theta + \Theta).
\end{aligned}
$$

Therefore, the amplitude ratio is $|\gamma|$ and the phase difference is Θ. Knowledge of this relation makes it easy to draw phasors Y_1 and X_1 as shown in Fig. 3.4(a).

If another oscillation quantity Z' is given by a sum of X' and Y', the complex amplitude Z_1 is simply expressed by a vector sum of X_1 and Y_1;

$$
Z_1 = X_1 + Y_1, \tag{3.22}
$$

as in Fig. 3.4(b), because the relation $Z' = X' + Y'$ is rewritten as

$$
\begin{aligned}
\mathrm{Re}[Z_1 e^{i\omega t}] &= \mathrm{Re}[X_1 e^{i\omega t}] + \mathrm{Re}[Y_1 e^{i\omega t}] \\
&= \mathrm{Re}[(X_1 + Y_1)e^{i\omega t}].
\end{aligned}
$$

In this way, one can actually see the phase relation among X', Y' and Z', before calculating it using trigonometric addition formulas.

Secondly, use of complex exponential functions often makes tedious calculation easy. The time derivative of $e^{i\omega t}$ is simply given by multiplying $i\omega$; $de^{i\omega t}/dt = i\omega e^{i\omega t}$. Therefore, the time derivative of $X' = \mathrm{Re}[X_1 e^{i\omega t}]$ is given by

$$
\frac{dX'}{dt} = \mathrm{Re}[i\omega X_1 e^{i\omega t}].
$$

The complex amplitude of the oscillations dX'/dt is given by the product of the complex amplitude X_1 of the original oscillation quantity with the factor $i\omega$.

In the same way, the oscillation quantity given by the time integral of X' is given by

$$\int X'dt = \text{Re}\left[\frac{X_1}{i\omega}e^{i\omega t}\right].$$

The complex amplitude is given by dividing the original complex amplitude X_1 with the factor $i\omega$.

Figure 3.5 represents the phasor diagram, where the phase relations among X_1, $i\omega X_1$, and $X_1/(i\omega)$ are shown. As the imaginary unit i is written as $e^{i\pi/2}$, multiplying/dividing X_1 with $i\omega$ corresponds to advancing/delaying the phase by $\pi/2$. Therefore, the phase lead of $i\omega X_1$ relative to X_1 is $\pi/2$, while the phase delay of $X_1/(i\omega)$ relative to X_1 is $\pi/2$.[1]

Before ending this section, consider the time average of a product of two oscillations $X' = A\cos(\omega t)$ and $Y' = B\cos(\omega t + \Theta)$, which is given by

$$\langle X'Y'\rangle_t = \frac{\omega}{2\pi}\int_0^{2\pi/\omega} X'Y'dt. \tag{3.23}$$

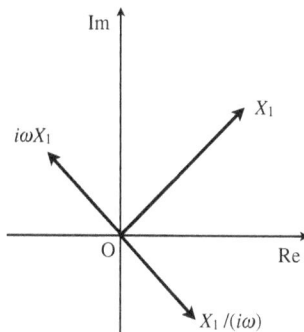

Figure 3.5: Phasor diagram.

---

[1]If X_1 represents the phasor of velocity, $i\omega X_1$ and $X_1/(i\omega)$ indicates those of acceleration and displacement. Their phase relations are apparent in Fig. 3.5.

Execution of the integral on the right hand side yields

$$\langle X'Y' \rangle_t = \frac{1}{2} AB \cos \Phi. \tag{3.24}$$

Therefore, when the complex amplitudes of X' and Y' are expressed as X_1 and Y_1, we can write down a mathematical formula

$$\langle X'Y' \rangle_t = \frac{1}{2} \mathrm{Re}[X_1^\dagger Y_1] \left(= \frac{1}{2} \mathrm{Re}[Y_1^\dagger X_1] \right), \tag{3.25}$$

where the symbol $\dagger$ on the right represents taking the complex conjugate. Specifically, for $X_1 = Ae^{i\theta}$, $X_1^\dagger$ is written as

$$X_1^\dagger = Ae^{-i\theta}. \tag{3.26}$$

This formula is used when we discuss the time-averaged energy or power of the oscillation system. If $X' = Y'$ holds, the time average becomes

$$\langle X'^2 \rangle_t = \frac{1}{2} |X_1|^2 = \frac{1}{2} A^2.$$

3.2.2 *Resonance curve and Q value of mechanical oscillation system*

Suppose an external periodic force f' with angular frequency ω is applied to a system shown in Fig. 3.1. The equation of motion is written as

$$M \frac{d^2 \xi'}{dt^2} + R \frac{d\xi'}{dt} + K\xi' = f'. \tag{3.27}$$

By introducing the complex amplitudes ξ_1 and f_1 for ξ' and f', respectively, we obtain

$$-\omega^2 M \xi_1 + i\omega R \xi_1 + K \xi_1 = f_1. \tag{3.28}$$

In the complex plane, the first term on the left hand side is directed oppositely to the third term, and the second term is directed normal to the first and the third terms, as shown in Fig. 3.6. The vector sum of these terms yields f_1. Since the complex amplitude ξ_1 is directed toward the third term, it is easy to find by eye the phasing between

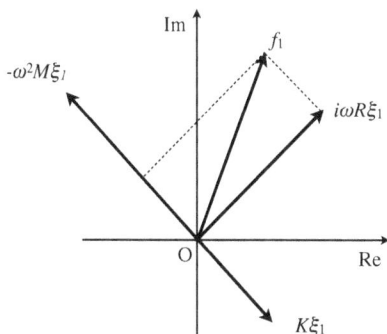

Figure 3.6: Phasor representation of the equation of motion.

ξ_1 and f_1. The solution is explicitly given by algebraic operation as

$$\xi_1 = \frac{f_1}{K - M\omega^2 + i\omega R}. \tag{3.29}$$

If we use the symbols $\tau_E = M/R$ and $\omega_0 = \sqrt{K/M}$ introduced in Section 3.1, ξ_1 is rewritten as

$$\xi_1 = \frac{f_1/M}{\omega_0^2 - \omega^2 + i\dfrac{\omega}{\tau_E}}. \tag{3.30}$$

The velocity u' is given by $u' = d\xi'/dt$, and therefore, its complex amplitude u_1 is expressed as $u_1 = i\omega\xi_1$;

$$u_1 = \frac{i\omega f_1/M}{\omega_0^2 - \omega^2 + i\dfrac{\omega}{\tau_E}} = \frac{if_1/M}{\omega_0^2/\omega - \omega + \dfrac{i}{\tau_E}}$$

$$= \frac{if_1}{\omega_0 M} \frac{1}{\left(\dfrac{\omega_0}{\omega} - \dfrac{\omega}{\omega_0}\right) + i\dfrac{1}{\omega_0 \tau_E}}.$$

Therefore,

$$|u_1|^2 = \left(\frac{|f_1|}{\omega_0 M}\right)^2 \frac{1}{\left(\dfrac{\omega_0}{\omega} - \dfrac{\omega}{\omega_0}\right)^2 + \left(\dfrac{1}{\omega_0 \tau_E}\right)^2}. \tag{3.31}$$

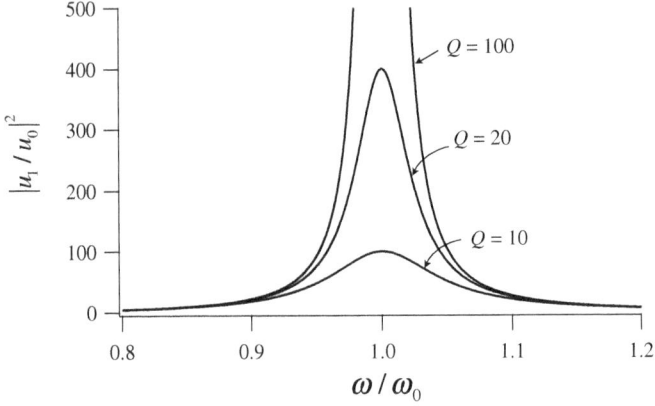

Figure 3.7: Resonance curve of $|u_1/u_0|^2$ vs. ω/ω_0, where $u_0 = f_1/(\omega_0 M)$. The graph shape changes with $Q = \omega_0 \tau_E$.

Figure 3.7 shows a resonance curve where $|u_1|^2$ is plotted against a normalized angular frequency, ω/ω_0. When $\omega/\omega_0 = 1$, namely when the external angular frequency is the same as the natural angular frequency of the undamped system, $|u_1|^2$ takes a maximum

$$|u_1|^2_{max} = \left(\frac{|f_1|}{\omega_0 M}\right)^2 \times (\omega_0 \tau_E)^2 = \left(\frac{|f_1|\tau_E}{M}\right)^2. \qquad (3.32)$$

We investigate here the peak width of the resonance curve in Fig. 3.7. For this purpose, we introduce a new variable ϵ given by

$$\omega/\omega_0 = 1 + \epsilon, \quad \epsilon \ll 1.$$

Since $\omega_0/\omega \approx 1 - \epsilon$, $|u_1|^2$ is simplified as

$$|u_1|^2 \approx \left(\frac{|f_1|}{\omega_0 M}\right)^2 \frac{1}{[1 - \epsilon - (1 + \epsilon)]^2 + \left(\dfrac{1}{\omega_0 \tau_E}\right)^2}$$

$$= \left(\frac{|f_1|}{\omega_0 M}\right)^2 \frac{1}{4\epsilon^2 + \left(\dfrac{1}{\omega_0 \tau_E}\right)^2}.$$

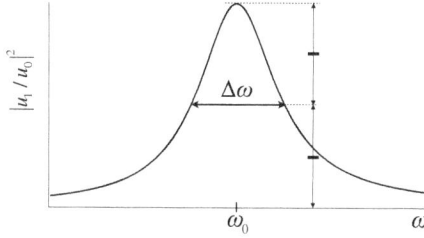

Figure 3.8: Resonance curve; $\Delta\omega$ represents the full width at half maximum.

It is found that the ω–dependence of $|u_1|^2$ is represented by a Lorentz curve, as presented in Fig. 3.8. We can see that $|u_1|^2$ becomes a half of $|u_1|^2_{max}$ when $4\epsilon^2$ is equal to $1/(\omega_0\tau_E)^2$, namely when

$$\epsilon = \pm\frac{1}{2\omega_0\tau_E}. \tag{3.33}$$

Therefore, the full width at half maximum of the curve is given by 2ϵ. In other words, an angular frequency difference $\Delta\omega = \omega_+ - \omega_-$ between $\omega_\pm = \omega_0 \pm 1/(2\tau_E)$ presents the width of the resonance curve as shown in Fig. 3.8. It should be noted that $\Delta\omega$ is related with ω_0 through τ_E:

$$\frac{\omega_0}{\Delta\omega} = \omega_0\tau_E. \tag{3.34}$$

The right-hand side is the same as the Q value obtained for the freely damped oscillation system. So we arrive at

$$Q = \frac{\omega_0}{\Delta\omega}, \tag{3.35}$$

which means that a high-Q system exhibits a sharp resonance.

Q value of the forced oscillation system can be derived directly through the definition $Q = \omega_0 E/W$ in Eq. (3.1). The time-averaged energy E is a sum of time-averaged kinetic energy and potential energy:

$$E = \frac{1}{2}M\langle u'^2\rangle_t + \frac{1}{2}K\langle\xi'^2\rangle_t = \frac{1}{4}M|u_1|^2 + \frac{1}{4}K|\xi_1|^2.$$

Substituting $\xi_1 = u_1/(i\omega)$ yields

$$E = \frac{1}{4}\left(M + \frac{K}{\omega^2}\right)|u_1|^2 = \frac{1}{4}M\left[1 + \left(\frac{\omega_0}{\omega}\right)^2\right]|u_1|^2. \qquad (3.36)$$

The time averaged power W supplied from the external force is given by

$$W = \langle f'u' \rangle_t = \frac{1}{2}\mathrm{Re}[f_1 u_1^\dagger].$$

If we substitute the solution of the equation of motion derived above for u_1, W is rewritten as follows:

$$W = \frac{1}{2}\mathrm{Re}\left[f_1 \frac{(-i)f_1^\dagger}{\omega_0 M} \frac{1}{\dfrac{\omega_0}{\omega} - \dfrac{\omega}{\omega_0} - \dfrac{i}{\omega_0 \tau_E}}\right]$$

$$= \frac{|f_1|^2}{2\omega_0 M}\mathrm{Re}\left[\frac{-i}{\dfrac{\omega_0}{\omega} - \dfrac{\omega}{\omega_0} - \dfrac{i}{\omega_0 \tau_E}}\right]$$

$$= \frac{|f_1|^2}{2\omega_0 M}\mathrm{Re}\left[\frac{-i\left(\dfrac{\omega_0}{\omega} - \dfrac{\omega}{\omega_0} + \dfrac{i}{\omega_0 \tau_E}\right)}{\left(\dfrac{\omega_0}{\omega} - \dfrac{\omega}{\omega_0}\right)^2 + \left(\dfrac{1}{\omega_0 \tau_E}\right)^2}\right]$$

$$= \frac{|f_1|^2}{2\omega_0 M}\frac{1}{\omega_0 \tau_E}\frac{1}{\left(\dfrac{\omega_0}{\omega} - \dfrac{\omega}{\omega_0}\right)^2 + \left(\dfrac{1}{\omega_0 \tau_E}\right)^2}$$

By recalling Eq. (3.31), W is further simplified as

$$W = \frac{|f_1|^2}{2\omega_0 M}\frac{1}{\omega_0 \tau_E}\frac{|u_1|^2}{\left(\dfrac{|f_1|^2}{\omega_0 M}\right)^2} = \frac{M|u_1|^2}{2\tau_E}. \qquad (3.37)$$

Finally, combining Eqs. (3.37) and (3.36) yields

$$Q = \frac{1}{2}\omega_0 \tau_E \left[1 + \left(\frac{\omega_0}{\omega}\right)^2\right].\tag{3.38}$$

Therefore, when ω equals to ω_0, Q is given by

$$Q = \omega_0 \tau_E.\tag{3.39}$$

Q-value can be obtained either from amplitude damping of freely decaying oscillations or from resonance curve of forced oscillations.

3.3 Acoustic Energy and Acoustic Intensity

In order to discuss the Q value of an acoustic resonance tube, consider the acoustic energy of the gas particle. Suppose that a gas particle in an acoustic resonance tube is subjected to pressure fluctuation p' and velocity fluctuation u'. In the case of a lossless resonance tube, the equation of motion and the equation of continuity read

$$\rho_m \frac{\partial u'}{\partial t} + \frac{\partial p'}{\partial x} = 0,\tag{3.40}$$

$$\frac{1}{c_S^2}\frac{\partial p'}{\partial t} + \rho_m \frac{\partial u'}{\partial x} = 0,\tag{3.41}$$

as we have seen in Chapter 2. By multiplying the equation of motion by u' and the equation of continuity by p'/ρ_m and adding them together, we have

$$\rho_m u' \frac{\partial u'}{\partial t} + \frac{1}{\rho_m c_S^2} p' \frac{\partial p'}{\partial t} + u'\frac{\partial p'}{\partial x} + p'\frac{\partial u'}{\partial x} = 0,$$

which is simplified as

$$\frac{\partial}{\partial t}\left(\frac{1}{2}\rho_m u'^2 + \frac{1}{2}\frac{p'^2}{\rho_m c_S^2}\right) + \frac{\partial}{\partial x}(p'u') = 0.\tag{3.42}$$

This equation represents a conservation law of acoustic energy. Namely, $\rho_m u'^2/2$ denotes the acoustic kinetic energy per unit volume, and $p'^2/(2\rho_m c_S^2)$ expresses the acoustic potential energy per unit volume; the sum of them means the *acoustic energy density*. The term $p'u'$ is *acoustic intensity*, which means the rate of energy transport per unit area. In this textbook, we specifically consider the time-averages of acoustic energy density and acoustic intensity:

$$e_d = \frac{1}{2}\rho_m \langle u'^2 \rangle_t + \frac{1}{2}\frac{\langle p'^2 \rangle_t}{\rho_m c_S^2}, \tag{3.43}$$

$$I = \langle p' u' \rangle_t \tag{3.44}$$

Thus, the energy E stored in the resonance tube is given by a volume integral of the acoustic energy density over the resonance tube:

$$E = \int_{\text{resonance tube}} e_d \, dV. \tag{3.45}$$

The acoustic power $\tilde{I}$ going through the cross section with cross-sectional area A is given by

$$\tilde{I} = AI. \tag{3.46}$$

3.4 Acoustic Resonance Tube without Dissipation

3.4.1 *Derivation of acoustic field*

As we have seen in Chapter 2, the acoustic pressure fluctuation p' obeys the wave equation

$$\frac{\partial^2 p'}{\partial t^2} - c_S^2 \frac{\partial^2 p'}{\partial x^2} = 0,$$

where c_S denotes the adiabatic speed of sound. Let's use this equation to derive the acoustic field in a lossless resonance tube. We assume a solution $p' = \text{Re}[p_1 e^{i\omega t}]$ for steady periodic oscillations of angular frequency ω. By inserting it into the wave equation, it is easy to see that the complex amplitude p_1 is written as

$$p_1 = C_+ e^{-ik_S x} + C_- e^{ik_S x} \tag{3.47}$$

where

$$k_S = \frac{\omega}{c_S} \tag{3.48}$$

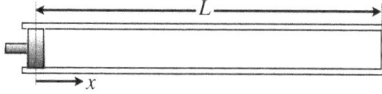

Figure 3.9: Acoustic resonance tube.

is adiabatic wavenumber; C_+ and C_- are complex constants. The first term on the right hand side means the complex amplitude of forward-traveling wave, and the second term represents that of backward-traveling wave, respectively. We should note that the unknown constants C_+ and C_- are determined by boundary conditions of the resonance tube, whereas k_S is derived from the thermodynamic relation between pressure and density.

Figure 3.9 shows a simple example of the acoustic resonance tube with one end closed and the other end plugged by a piston that oscillates back and forth with constant velocity amplitude U_0 and angular frequency ω. The origin, $x = 0$, of axial coordinate is taken at the piston's average position, and the right end of the resonance tube is set to $x = L$. The boundary conditions are given for the acoustic particle velocity $u' = \mathrm{Re}[u_1 e^{i\omega t}]$ as

$$u_1|_{x=0} = U_0 \tag{3.49}$$

$$u_1|_{x=L} = 0 \tag{3.50}$$

From the equation of motion

$$\rho_m \frac{\partial u'}{\partial t} = -\frac{\partial p'}{\partial x},$$

we see that the complex amplitude u_1 is related to p_1 as

$$u_1 = \frac{i}{\omega \rho_m} \frac{dp_1}{dx}. \tag{3.51}$$

Inserting Eqs. (3.47) into (3.51) yields

$$u_1 = \frac{C_+}{z_S} e^{-ik_S x} - \frac{C_-}{z_S} e^{ik_S x} \tag{3.52}$$

where

$$z_S = \rho_m c_S \tag{3.53}$$

is characteristic impedance of the gas. The boundary conditions in Eq. (3.50) are then expressed as

$$\frac{C_+}{z_S} - \frac{C_-}{z_S} = U_0 \tag{3.54}$$

$$\frac{C_+}{z_S}e^{-ik_S L} - \frac{C_-}{z_S}e^{ik_S L} = 0. \tag{3.55}$$

Solving the coupled equations above gives

$$C_+ = \frac{z_S U_0 e^{ik_S L}}{2i \sin(k_S L)} \tag{3.56}$$

$$C_- = \frac{z_S U_0 e^{-ik_S L}}{2i \sin(k_S L)} \tag{3.57}$$

when $k_S \neq n\pi$ ($n = 1, 2, \ldots$). Finally, by inserting Eqs. (3.56) and (3.57) into Eqs. (3.47) and (3.52), we obtain

$$p_1(x) = -iz_S \frac{\cos[k_S(L-x)]}{\sin(k_S L)} U_0 \tag{3.58}$$

$$u_1(x) = \frac{\sin[k_S(L-x)]}{\sin(k_S L)} U_0 \tag{3.59}$$

Because both p_1 and u_1 have a factor of $\sin(k_S L)$ in denominator, the amplitudes $|p_1|$ and $|u_1|$ diverge when $k_S = n\pi$ ($n = 1, 2, \ldots$). In other words, acoustic resonance occurs when $k_S = n\pi$. Also, they increase in proportion to the piston velocity U_0. So U_0 controls the magnitude of acoustic oscillations.

Figures 3.10 (a1) and (a2) show the axial distributions of $|p_1|$ and $|u_1|$, respectively when $k_S L = 1.05\pi$ and 2.1π. A pressure antinode (velocity node) is created at the closed end of the resonance tube, $x = L$, where $|p_1|$ becomes the maximum. A velocity antinode (pressure node) is separated from $x = L$ by a quarter wavelength,

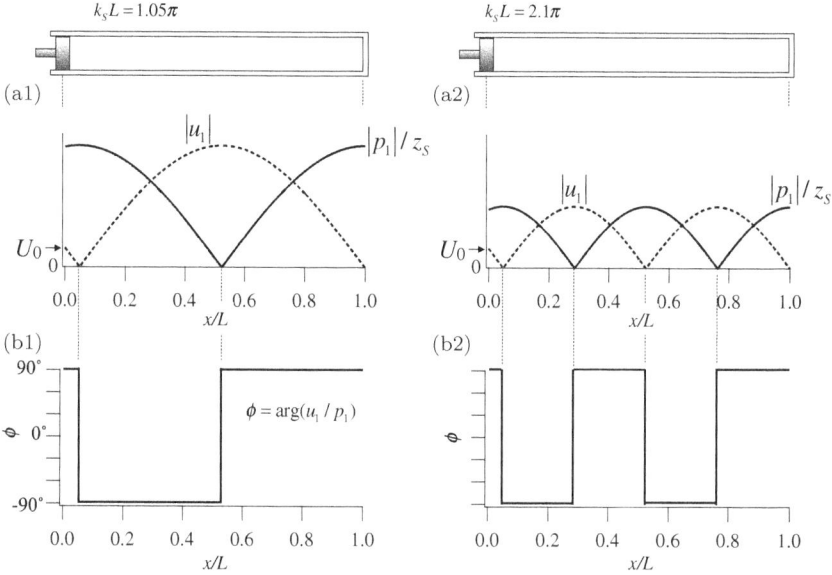

Figure 3.10: Acoustic fields when $k_S L = 1.05\pi$ (a1, b1) and 2.1π (a2, b2).

where $|u_1|$ becomes the maximum. The ratio of maxima of $|p_1|$ to $|u_1|$ is equal to the characteristic acoustic impedance z_S of the gas.

Figures 3.10 (b1) and (b2) present a phase lead ϕ of p_1 relative to u_1, respectively when $k_S L = 1.05\pi$ and 2.1π. The phase lead ϕ is a step function of x, changing from $-\pi/2$ to $\pi/2$ at every antinode and node.

Pressure oscillations p' and velocity oscillations u' are respectively written as a function of time t;

$$p' = z_S \frac{\cos[k_S(L-x)]}{\sin(k_S L)} U_0 \sin(\omega t)$$

$$u' = \frac{\sin[k_S(L-x)]}{\sin(k_S L)} U_0 \cos(\omega t).$$

As the variables x and t are separated from each other, the axial positions of pressure antinodes and velocity antinodes do not change with time. Such a wave is called a standing wave.

3.4.2 *Quality factor of resonance tube*

When a gas particle oscillates adiabatically in a resonance tube, the time-averaged acoustic energy density is expressed by

$$e_d = \frac{\langle p'^2 \rangle_t}{2\rho_m c_S^2} + \frac{1}{2}\rho_m \langle u'^2 \rangle_t.$$

For steady periodic oscillations, e_d is given by

$$e_d = \frac{|p_1|^2}{4\rho_m c_S^2} + \frac{1}{4}\rho_m |u_1|^2. \tag{3.60}$$

By substituting Eqs. (3.58) and (3.59), e_d is simplified as

$$e_d = \frac{\rho_m U_0^2}{4\sin^2(k_S L)}. \tag{3.61}$$

One can notice that e_d is uniform throughout the resonance tube, as it is independent of x. So the volume integral in Eq. (3.45) is executed only by multiplying e_d with volume V of the resonance tube. Therefore, the acoustic energy E stored in the resonator is given by

$$E = \frac{\rho_m U_0^2}{4\sin^2(k_S L)}V. \tag{3.62}$$

Since the denominator contains a factor of $\sin(k_S L)$, E becomes infinitely large when $k_S L$ equals $n\pi$ $(n = 1, 2, \ldots)$, namely when the resonance occurs, as shown in Fig. 3.11.

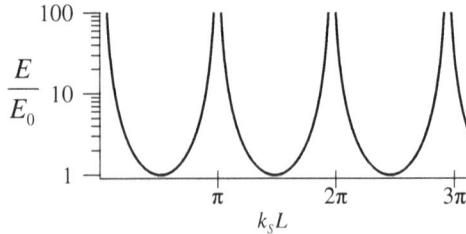

Figure 3.11: Acoustic energy E as a function of $k_S L$. The ratio to $E_0 = \rho_m V U_0^2/4$ is shown.

The time-average of the work done by the piston per unit time is given by

$$W = \frac{A}{2}\text{Re}[P_0 U_0^\dagger] \tag{3.63}$$

where A represents the cross-sectional area of the piston surface, and P_0 means the force per unit area of the piston surface. More specifically, P_0 is given by $P_0 = p_1(0)$ and therefore,

$$P_0 = -iz_S \frac{\cos(k_S L)}{\sin(k_S L)} U_0. \tag{3.64}$$

Because P_0 contains the imaginary unit i, P_0 is out of phase by $\pi/2$ from U_0. Therefore, the time average of the work W is zero while E is finite. As a result, $Q = \omega E/W$ goes infinitely large:

$$\frac{1}{Q} = 0.$$

Such an unrealistic result is obtained because we have ignored the viscosity and thermal conductivity of the gas and the associated energy losses.

3.5 Viscous Loss in a Resonance Tube

3.5.1 *Equation of motion for viscous fluid and its solution*

Energy dissipations due to the viscosity and thermal conductivity of the gas are unavoidable in an actual acoustic resonance tube. In this section, the energy dissipation due to the viscosity is described, and then the quality factor of a realistic resonance tube is discussed using the complex wavenumber k and complex characteristic impedance z for sound waves in the tube.

The equation of motion for viscous fluid is given by

$$\rho_m \frac{\partial u'}{\partial t} = -\frac{\partial p'}{\partial x} + \mu \triangle_\perp u' \tag{3.65}$$

where μ stands for the viscosity of the gas ($\mu = \rho_m \nu$); $\triangle_\perp$ represents Laplace's operator with respect to the cross-section of the tube. For a circular tube, $\triangle_\perp$ is expressed by

$$\triangle_\perp = \frac{\partial^2}{\partial r^2} + \frac{1}{r}\frac{\partial}{\partial r}$$

where r denotes the radial coordinate in a circular coordinate system. In order to find a solution for a steady periodic state with angular frequency ω, we use the complex notation for the axial particle velocity u' and acoustic pressure p', and insert $u' = \mathrm{Re}[u_1 e^{i\omega t}]$ and $p' = \mathrm{Re}[p_1 e^{i\omega t}]$ into Eq. (3.65) to have

$$u_1 = \frac{i}{\omega\rho_m}\frac{dp_1}{dx} - i\frac{\nu}{\omega}\triangle_\perp u_1, \tag{3.66}$$

We adopt a nonslip boundary condition at the wall surface. Therefore,

$$u_1|_{r=r_0} = 0. \tag{3.67}$$

Also, to ensure the smooth velocity profile on the central axis ($r = 0$), we set

$$\left.\frac{\partial u_1}{\partial r}\right|_{r=0} = 0. \tag{3.68}$$

3.5.2   *Solution of equation of motion*

For an inviscid gas, the velocity has a solution in Eq. (3.51)

$$u_1 = \frac{i}{\omega\rho_m}\frac{dp_1}{dx}.$$

So, we assume a solution of the form [16]

$$u_1 = \frac{i}{\omega\rho_m}\frac{dp_1}{dx}(1 - f_\nu) \tag{3.69}$$

where f_ν is an unknown complex function of r. The factor $1 - f_\nu$ represents the impact of the viscosity of the gas. Inserting u_1 into Eq. (3.66) gives

$$f_\nu - \frac{\nu}{i\omega}\Delta_\perp f_\nu = 0. \tag{3.70}$$

We introduce a characteristic transverse length δ_ν given by

$$\delta_\nu = \sqrt{\frac{2\nu}{\omega}}. \tag{3.71}$$

By using a new variable

$$\eta_\nu = (i - 1)\frac{r}{\delta_\nu}, \tag{3.72}$$

the operators in $\Delta_\perp$ are transformed to

$$\frac{1}{r}\frac{\partial}{\partial r} = -\frac{2i}{\eta_\nu \delta_\nu^2}\frac{\partial}{\partial \eta_\nu}$$

$$\frac{\partial^2}{\partial r^2} = -\frac{2i}{\delta_\nu^2}\frac{\partial^2}{\partial \eta_\nu^2}.$$

Therefore, after some algebraic calculations, Eq. (3.70) reduces to

$$\frac{\partial^2 f_\nu}{\partial \eta_\nu^2} + \frac{1}{\eta_\nu}\frac{\partial f_\nu}{\partial \eta_\nu} + f_\nu = 0.$$

This equation is equivalent to the Bessel's differential equation

$$\frac{\partial^2 y}{\partial \eta^2} + \frac{1}{\eta}\frac{\partial y}{\partial \eta} + \left(1 - \frac{n^2}{\eta^2}\right)y = 0$$

with $n = 0$. The fundamental solutions are the Bessel fucntion $J_0(\eta)$ of the 0-th order and the Neumann function $N_0(\eta)$ of the 0-th order.

Therefore, the general solution is written as

$$f_\nu = C_1 J_0(\eta_\nu) + C_2 N_0(\eta_\nu).$$

Unknown constants C_1 and C_2 are determined by the boundary conditions

$$f_\nu|_{r=r_0} = 1 \tag{3.73}$$

$$\left.\frac{\partial f_\nu}{\partial r}\right|_{r=0} = 0 \tag{3.74}$$

Because N_0 changes as a logarithmic function of η as $\eta \to 0$, C_2 should be zero to satisfy the boundary condition in Eq. (3.74). The unknown constant C_1 is determined by Eq. (3.73). Finally, f_ν is given by

$$f_\nu = \frac{J_0(\eta_\nu)}{J_0(\eta_{\nu0})}, \tag{3.75}$$

with

$$\eta_{\nu0} = (i-1)\frac{r_0}{\delta_\nu}. \tag{3.76}$$

3.5.3   *Graphical representation of velocity fluctuation*

The function f_ν, responsible for the radial profile of complex velocity amplitude u_1, has a parameter r_0/δ_ν. This ratio is related to the viscous relaxation time τ_ν in the cross-section of the tube

$$\tau_\nu = \frac{r_0^2}{2\nu} \tag{3.77}$$

through

$$\omega\tau_\nu = \left(\frac{r_0}{\delta_\nu}\right)^2. \tag{3.78}$$

Figure 3.12 represents the magnitude and phase of a radial distribution function $1 - f_\nu$ of u_1 for representative values of $\omega\tau_\nu$. The velocity amplitude goes to zero in the vicinity of the tube wall at $r = r_0$, as a result of the viscous interaction between the gas and the wall. When $\omega\tau_\nu \gg 1$, the velocity magnitude shows a peak at

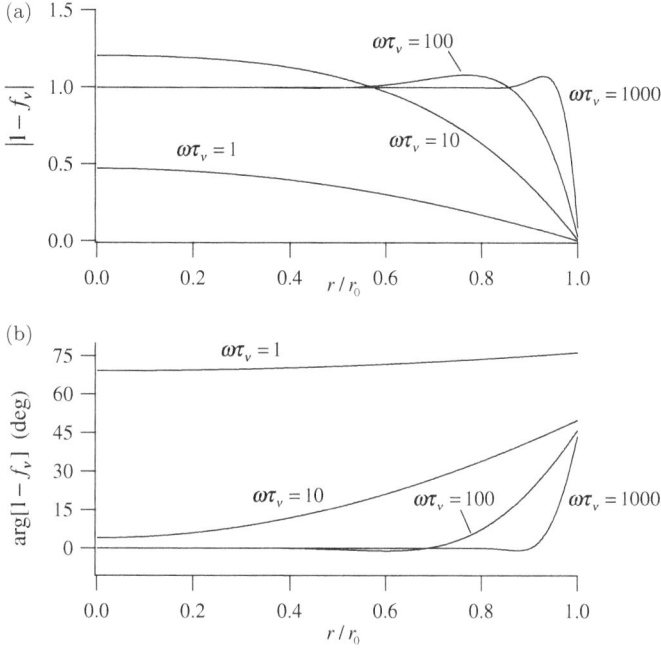

Figure 3.12: Radial distribution of velocity amplitude u_1 of viscous fluid: (a) magnitude and (b) phase.

about $2\delta_\nu$ separated from the wall, while it becomes uniform in a core region more than $3\delta_\nu$ away from the wall. In other words, the influence of the wall extends for about $3\delta_\nu$ from the wall. The phase of u_1 decreases as r/r_0 decreases from 1, while it shows a flat profile in the core region in the same way as the magnitude. These radial profiles have been confirmed by recent optical experiments.

3.5.4 *Viscous energy dissipation rate*

The gas parcel moves with velocity u' against viscous force $\mu\triangle_\perp u'$ in the cylindrical flow channel. So a product of u' and $\mu\triangle_\perp u'$ represents the viscous energy dissipation rate per unit time. The time average is given by

$$\left\langle u'(\mu\triangle_\perp u')\right\rangle_t = \left\langle u'\left(\rho_m \frac{\partial u'}{\partial t} + \frac{\partial p'}{\partial x}\right)\right\rangle_t = \left\langle u'\left(\frac{\partial p'}{\partial x}\right)\right\rangle_t.$$

Note that the relation $\langle u' \partial u'/\partial t \rangle = 0$ holds because u' and $\partial u'/\partial t$ oscillate out of phase by $90°$ from each other. By the mathematical formula of Eq. (3.25), we can simplify the expression as

$$\left\langle u'\left(\frac{\partial p'}{\partial x}\right)\right\rangle_t = \frac{1}{2}\mathrm{Re}\left[\left(\frac{i}{\omega\rho_m}\frac{dp_1}{dx}(1-f_\nu)\right)^{\dagger}\frac{dp_1}{dx}\right]$$

$$= \frac{1}{2\omega\rho_m}\mathrm{Re}\left[\left(i(1-f_\nu)\frac{dp_1}{dx}\right)^{\dagger}\frac{dp_1}{dx}\right]$$

$$= \frac{1}{2\omega\rho_m}\left|\frac{dp_1}{dx}\right|^2\mathrm{Re}\left[-i(1-f_\nu^{\dagger})\right].$$

If we set $f_\nu = \mathrm{Re}f_\nu + i\mathrm{Im}f_\nu$,

$$-i(1 - f_\nu^{\dagger}) = -i(1 - \mathrm{Re}f_\nu + i\mathrm{Im}f_\nu) = \mathrm{Im}f_\nu + i(\mathrm{Re}f_\nu - 1).$$

Finally we see that the rate of energy dissipation due to viscosity per unit time is given by the following form:

$$\langle u'(\mu\triangle_{\perp}u')\rangle_t = \frac{1}{2\omega\rho_m}\left|\frac{dp_1}{dx}\right|^2\mathrm{Im}f_\nu. \qquad (3.79)$$

Figure 3.13 shows the radial distribution of $\mathrm{Im}\,f_\nu$ for cylindrical flow channels with several $\omega\tau_\nu$ values. Negative $\mathrm{Im}\,f_\nu$ means dissipation of energy, while positive value corresponds to production. It is seen that

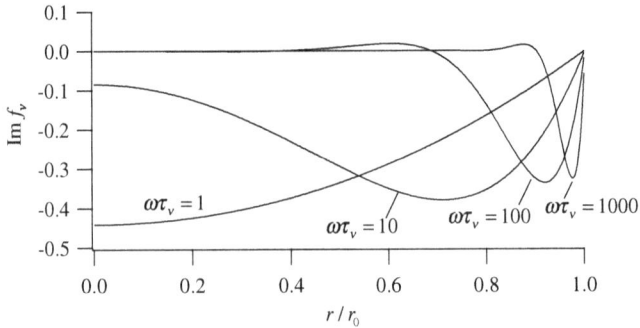

Figure 3.13: Radial distribution of the rate of energy dissipation due to viscosity per unit time.

the viscous energy dissipation becomes the greatest at the positions about δ_ν away from the channel walls, where the velocity amplitude greatly changes as shown in Fig. 3.12. It will be shown in Chapter 6 that the cross-sectional mean of viscous energy dissipation rate per unit time, $\langle\langle u'(\mu\triangle_\perp u')\rangle_t\rangle_r$, represents work source W_ν.

3.5.5 *Cross-sectional mean velocity*

Consider here the cross-sectional mean of acoustic particle velocity, $\langle u'\rangle_r = \mathrm{Re}[u_{1r}e^{i\omega t}]$. The complex amplitude u_{1r} of the cross-sectional mean velocity is given from $u_{1r} = \langle u_1\rangle_r$ as

$$u_{1r} = \frac{i}{\omega\rho_m}\frac{dp_1}{dx}(1-\chi_\nu) \tag{3.80}$$

where

$$\chi_\nu = \langle f_\nu\rangle_r. \tag{3.81}$$

Therefore, the relation between u_1 and u_{1r} is expressed by

$$u_1 = \frac{1-f_\nu}{1-\chi_\nu}u_{1r}. \tag{3.82}$$

Function χ_ν is given by using the Bessel function of the first kind, J_1, as

$$\chi_\nu = \frac{2J_1(\eta_{\nu 0})}{\eta_{\nu 0}J_0(\eta_{\nu 0})}. \tag{3.83}$$

The derivation is as follows. The explicit form of $\chi_\nu = \langle f_\nu\rangle_r$ is

$$\chi_\nu = \frac{1}{\pi r_0^2}\int_0^{r_0} 2\pi r f_\nu(r)dr. \tag{3.84}$$

By using the change of variable introduced in Eq. (3.72), we have

$$\chi_\nu = \frac{2}{\eta_{\nu 0}^2 J_0(\eta_{\nu 0})}\int_0^{\eta_{\nu 0}} \eta_\nu J_0(\eta_\nu)d\eta_\nu. \tag{3.85}$$

Use of a mathematical formula of the Bessel functions

$$\int_0^{\eta_{\nu 0}} \eta_\nu J_0(\eta_\nu)d\eta_\nu = \eta_{\nu 0} J_1(\eta_{\nu 0}) \tag{3.86}$$

yields the above expression of χ_ν.

The asymptotic forms of χ_ν are given by

$$\chi_\nu = \begin{cases} 1 - \dfrac{1}{12}(\omega\tau_\nu)^2 - i\dfrac{\omega\tau_\nu}{4} & (\omega\tau_\nu \ll \pi) \\ \sqrt{\dfrac{2}{\omega\tau_\nu}} \exp\left(-i\dfrac{\pi}{4}\right) & (\omega\tau_\nu \gg \pi). \end{cases} \tag{3.87}$$

Figure 3.14 shows (a) χ_ν, (b) the magnitude and (c) phase of $1 - \chi_\nu$ as a function of $\omega\tau_\nu$. In the region with $\omega\tau_\nu \gg 1$, $1 - \chi_\nu$ approaches 1. Therefore, u_{1r} should become the same as the velocity of the inviscid gas in Eq. (3.51). In the region with $\omega\tau_\nu \ll 1$, the amplitude of u_{1r} decreases to zero, while the phase u_{1r} advances by 90°.

3.6 Lossy Acoustic Resonance Tube

3.6.1 *Derivation of acoustic field*

3.6.1.1 *Wavenumver*

We have seen in Chapter 2 that the wavenumber k in a tube differs from the adiabatic wavenumber k_S because of viscosity and thermal conductivity of the gas. For the sound waves propagating through a cylindrical tube, k is expressed as

$$\frac{k}{k_S} = \sqrt{\frac{1 + (\gamma - 1)\chi_\alpha}{1 - \chi_\nu}} \tag{3.88}$$

where χ_α is a function of $\omega\tau_\alpha$ given by replacing kinematic viscosity ν with thermal diffusivity α in χ_ν [Eq. (3.85)]. See Chapter 5 for derivation of χ_α and k.

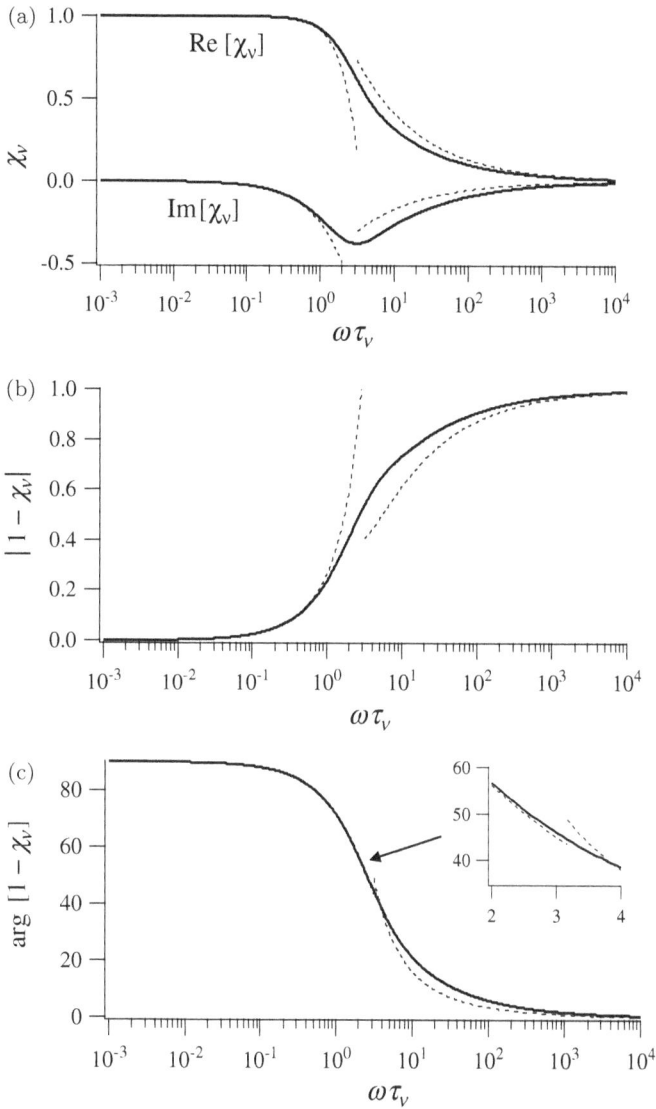

Figure 3.14: $\omega\tau_\nu$ dependence of χ_ν and $1 - \chi_\nu$: (a) real and imaginary parts of χ_ν, magnitude $|1 - \chi_\nu|$, (b) magnitude $|1 - \chi_\nu|$, and (c) phase $\arg[1 - \chi_\nu]$. Dashed curves indicates those calculated by using asymptotic forms.

3.6.1.2   *Characteristic acoustic impedance*

We have seen in Chapter 2 that the ratio of pressure to velocity
in adiabatic traveling waves is equal to characteristic acoustic
impedance $z_S = \rho_m c_S$ for the waves going in a positive direction. This
relation is modified by viscosity and thermal conductivity of the gas,
as we show below. By substituting the complex pressure amplitude
$p_1 = C_+ e^{-ikx}$ of the forward traveling wave into Eq. (3.80), we have

$$u_{1r} = \frac{k(1 - \chi_\nu)}{\omega \rho_m} p_1.$$

Therefore, the ratio, $z = p_1/u_{1r}$, of pressure to cross-sectional mean
velocity is given by

$$z = \frac{\omega \rho_m}{k(1 - \chi_\nu)}.$$

By further using Eq. (3.88), z is given by

$$\frac{z}{z_S} = \frac{1}{(1 - \chi_\nu)\sqrt{\dfrac{1 + (\gamma - 1)\chi_\alpha}{1 - \chi_\nu}}}. \tag{3.89}$$

As shown in Fig. 3.15, z approaches z_S in the limit of $\omega \tau_\nu \to \infty$,
whereas the magnitude of k increases and the phase goes to $45°$ as
$\omega \tau_\nu \to 0$.

3.6.1.3   *Decomposition into forward- and backward-traveling waves*

We have seen in the case of adiabatic plane sound waves that the
pressure oscillations can be expressed as a sum of forward- and
backward-traveling waves. The same technique can be applied to the
sound waves propagating in a tube by replacing k_S and z_S in Eqs.
(3.47) and (3.52) with k and z, respectively. Namely, we have

$$p_1 = C_+ e^{-ikx} + C_- e^{ikx} \tag{3.90}$$

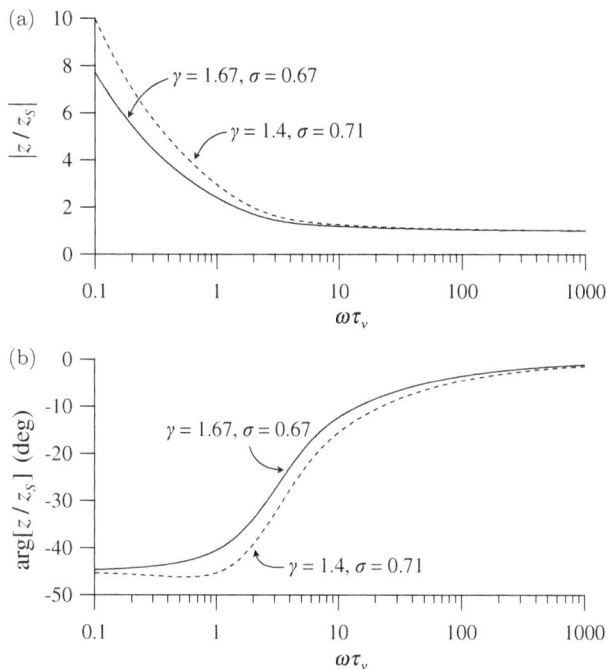

Figure 3.15: $\omega\tau_\nu$ dependence of z: (a) magnitude $|z/z_S|$ and (b) phase of z/z_S, $\arg[z/z_S]$.

and

$$u_{1r} = \frac{C_+}{z}e^{-ikx} - \frac{C_-}{z}e^{ikx}. \tag{3.91}$$

Unknown constants C_+ and C_- are determined by boundary conditions of a lossy resonance tube under consideration. When the resonance tube is closed at one end and driven at the other end by a piston as shown in Fig. 3.9, pressure and velocity oscillations are written as

$$p_1(x) = -iz\frac{\cos[k(L-x)]}{\sin(kL)}U_0 \tag{3.92}$$

$$u_{1r}(x) = \frac{\sin[k(L-x)]}{\sin(kL)}U_0. \tag{3.93}$$

3.6.2 Q value of lossy resonance tube

Let us try to derive Q value of a relatively wide resonance tube with $\omega\tau_\nu \gg 1$. The wavenumber k can be approximately written by

$$\frac{k}{k_S} = 1 + \frac{1-i}{2\sqrt{\omega\tau_\nu}} + \frac{(1-i)(\gamma-1)}{2\sqrt{\omega\tau_\alpha}},$$

as we have seen in Section 2.4 (see Chapter 5 for derivation). Note here that we have used a relation $\omega\tau_\nu = \omega\tau_\alpha/\sigma$, where σ denotes the Prandtl number of the gas. If we set

$$\zeta = \frac{1}{2\sqrt{\omega\tau_\nu}} + \frac{(\gamma-1)}{2\sqrt{\omega\tau_\alpha}},$$

k/k_S is expressed by $k/k_S = 1 + \zeta - i\zeta$. When $\omega\tau_\nu \gg 1$, it is further simplified as

$$\frac{k}{k_S} = 1 - i\zeta. \tag{3.94}$$

When $\omega\tau_\nu \gg 1$, the characteristic acoustic impedance z is approximated by

$$\frac{z}{z_S} = \frac{1}{1-\chi_\nu}\frac{1}{1-i\zeta} \approx (1+\chi_\nu)(1+i\zeta) \approx 1 + i\left(\zeta - \frac{1}{\sqrt{\omega\tau_\nu}}\right), \tag{3.95}$$

where we have used the asymptotic form of χ_ν

$$\chi_\nu = \sqrt{\frac{2}{\omega\tau_\nu}}\exp\left(-i\frac{\pi}{4}\right).$$

These approximations of k and z are used for derivation of Q in the following way.

When the acoustic resonance tube shown in Fig. 3.9 is driven by the piston at the end with constant velocity amplitude U_0 and angular frequency ω, the acoustic energy stored in the tube is given by replacing k_S with k in Eq. (3.62) as

$$E = \frac{\rho_m U_0^2}{4\sin^2(kL)}V. \tag{3.96}$$

The acoustic pressure P_0 at the piston end is given by

$$P_0 = -iz\frac{\cos kL}{\sin kL}U_0.$$

Therefore, the work done by the piston per unit time is expressed by

$$W = \frac{A}{2}U_0^2 \text{Re}\left[-iz\frac{\cos kL}{\sin kL}\right]. \tag{3.97}$$

In the case of the lossless resonance tube, the acoustic resonance takes place with

$$k_S L = n\pi, \tag{3.98}$$

where n is integer. When length L satisfies the condition above, $\cos kL$ and $\sin kL$ are transformed as follows.

$$\cos kL = \cos(k_S L - i\zeta k_S L)$$
$$= \frac{1}{2}\{\exp\left[i(k_S L - i\zeta k_S L)\right] + \exp\left[-i(k_S L - i\zeta k_S L)\right]\}$$
$$= \frac{1}{2}[(\cos k_S L + i\sin k_S L)e^{\zeta k_S L}$$
$$+(\cos k_S L - i\sin k_S L)e^{-\zeta k_S L}]$$
$$\approx \frac{1}{2}(1 + \zeta k_S L + 1 - \zeta k_S L) = 1$$
$$\sin kL = \sin(k_S L - i\zeta k_S L)$$
$$= \frac{1}{2i}\{\exp\left[i(k_S L - i\zeta k_S L)\right] - \exp\left[-i(k_S L - i\zeta k_S L)\right]\}$$
$$= \frac{1}{2i}[(\cos k_S L + i\sin k_S L)e^{\zeta k_S L}$$
$$-(\cos k_S L - i\sin k_S L)e^{-\zeta k_S L}]$$
$$\approx \frac{1}{2i}(1 + \zeta k_S L - 1 + \zeta k_S L) = -i\zeta k_S L$$

Note we have used relations $\cos k_S L = 1$ and $\sin k_S L = 0$, as well as $\zeta \ll 1$, during the transformation above. As a result, the acoustic energy E is obtained as

$$E = \frac{\rho_m U_0^2}{4(\zeta k_S L)^2} V.$$ (3.99)

Also, the work per unit time is given by

$$W = \frac{A}{2} U_0^2 \frac{z_S}{\zeta k_S L}.$$ (3.100)

Therefore, we obtain the Q as

$$Q = \frac{\omega E}{W} = \frac{\omega \rho_m}{2\zeta z_S k_S},$$

which can be simplified as

$$Q = \frac{1}{2\zeta}.$$ (3.101)

More specifically, Q of the resonance tube is expressed by [50]

$$\frac{1}{Q} = \frac{1}{\sqrt{\omega \tau_\nu}} + \frac{\gamma - 1}{\sqrt{\omega \tau_\alpha}}.$$ (3.102)

The Q value can be derived from the resonance curve, as we did in the case of the damped mechanical oscillator. For this purpose, we focus on the complex amplitude of pressure. By replacing k_S and z_S with k and z, $p_1(x)$ of the lossless case is transformed to that of the lossy case;

$$p_1(x) = -iz \frac{\cos[k(L - x)]}{\sin kL} U_0.$$ (3.103)

Let's consider the resonance curve by plotting the closed-end pressure at $x = L$, which is given by

$$p_e = p_1(L) = -iz \frac{1}{\sin kL} U_0.$$

If we change the length of the resonance tube from L to $L+l$ ($l \ll L$), we have

$$p_e = -iz \frac{1}{\sin k(L + l)} U_0.$$

Here, $\sin k(L + l)$ is rewritten as

$$\sin k(L + l) = \sin\left[(k_S - i\zeta k_S)(L + l)\right]$$

$$= \frac{1}{2i}\left[e^{\zeta k_S(L+l)}\exp[ik_S(L + l)]\right.$$

$$\left. -e^{-\zeta k_S(L+l)}\exp[-ik_S(L + l)]\right].$$

To further simplify the expression above, we use approximate equations

$$e^{\zeta k_S(L+l)} \approx 1 + \zeta k_S(L + l) \approx 1 + \zeta k_S L,$$

$$e^{-\zeta k_S(L+l)} \approx 1 - \zeta k_S(L + l) \approx 1 - \zeta k_S L,$$

and

$$\cos k_S(L + l) \approx \pm 1$$

$$\sin k_S(L + l) \approx \pm k_S l.$$

(Note that plus sign corresponds to even n in Eq. (3.98), and the minus sign to odd n.) Finally, we obtain

$$\sin k(L + l) \approx \pm\frac{1}{2i}\left[(1 + \zeta k_S L)(\pm 1 \pm ik_S l) - (1 - \zeta k_S L)(\pm 1 \mp ik_S l)\right]$$

$$= \pm\frac{k_S}{i}(\zeta L + il).$$

Therefore, the end pressure is given by

$$p_e = \pm\frac{z_S}{k_S(\zeta L + il)}U_0, \tag{3.104}$$

and so

$$|p_e|^2 = \frac{\left(\dfrac{z_S U_0}{k_S}\right)^2}{(\zeta L)^2 + l^2}.$$

When the resonance tube length is varied, $|p_e|^2$ shows a resonance curve given by a Lorentz curve, as in the case of the damped mechanical oscillator in Fig. 3.8. The curve of $|p_e|^2$ takes a maximum value with $l = 0$, and becomes half when $l = \zeta L$. Therefore, the full width at half maximum (FWHM) is $2\zeta L$. The ratio of the resonance tube length to FWHM is expressed by

$$\frac{L}{2\zeta L} = \frac{1}{2\zeta}.$$

As we see from Eq. (3.101), it is equal to Q of the resonance tube. Thus, the Q is obtainable from the resonance curve when the tube length is varied.

3.6.3 *Experimental acoustic field of resonance tube*

An acoustic resonance tube shown in Fig. 3.9 was built using a 1.04-m long cylindrical tube of 10.5 mm inner diameter [51]. The tube was filled with air at ambient temperature and pressure. One end of the tube was closed by a rigid plate, while the other was connected to an acoustic driver. The driving frequency was tuned to 148.5 Hz, equal to the fundamental natural frequency f_0 of the resonance tube. These experimental conditions yield $\omega \tau_\nu = 3.4 \times 10^3$. Taking into account that the Prandtl number of air is 0.71, $\omega \tau_\alpha$ is estimated as 2.4×10^3.

Figure 3.16 shows the acoustic field when the pressure amplitude was 782 Pa at the closed end: (a) the amplitudes $|p_1|$ of pressure and $|u_{1r}|$ of cross-sectional average velocity, (b) the phase lead of the cross-sectional average velocity relative to pressure, and (c) the acoustic intensity I. The cross-sectional average velocity u_{1r} was obtained from the velocity $u_1(0)$ measured on the central axis of the tube as

$$u_{1r} = \frac{1 - \chi_\nu}{1 - f_\nu(0)} u_1(0). \tag{3.105}$$

Equation (3.105) is derived from Eq. (3.82).

Since driving frequency was chosen to be the fundamental frequency of the resonance tube, the excited oscillating mode has the pressure maxima at ends, and the velocity maximum at the middle, as expected for adiabatic acoustic waves free from the effects of viscosity and thermal conductivity. The acoustic intensity I is, however, always finite throughout the resonance tube. Therefore, the

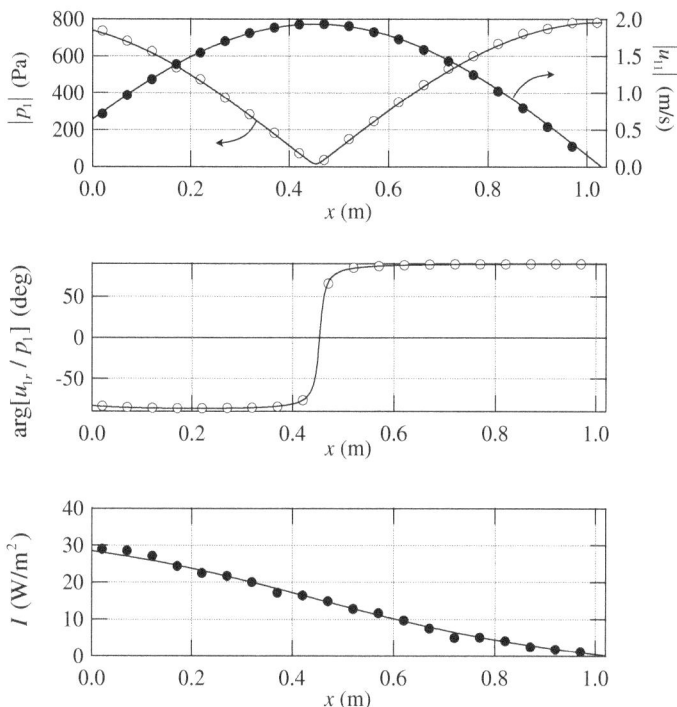

Figure 3.16: Experimental acoustic field of the resonance tube. The curve represents theoretical results.

non-zero velocity is realized at the velocity node in the middle of the resonance tube, and the phase difference between the pressure and the velocity is always different from $\pm\pi/2$, although the difference is small. The positive sign of I represents that it flows in the positive direction of x. So the acoustic driver on the left end of the resonance tube plays a role of sound source. The intensity I decreases as it flows down, owing to the energy dissipation caused by the viscous and thermal interactions between the gas and the internal surface of the tube. The intensity I becomes zero at the right end as a natural consequence that the acoustic waves do not go out of the resonance tube. We should note that I goes to zero with keeping its negative slope at the closed end because of the energy dissipation due to the non-zero pressure amplitude. The curves in the figures show the theoretical results of the acoustic theory [Eqs. (3.92) and (3.93)], which quantitatively is in good agreement with the experimental data.

Now let's consider the quality factor Q of the resonance tube. We see that Eqs. (3.58) and (3.59) are reasonable approximations to the pressure and velocity amplitudes. So we can evaluate the acoustic energy stored E in the resonance tube by using Eq. (3.62). The acoustic power W necessary to sustain the steady oscillations is derived from the product of the acoustic intensity I at the left end and the cross sectional area A of the tube. The Q value determined from the stored energy and the supplied power thus derived turned out to be $Q = 36$. The theoretical quality factor of the resonance tube is approximately given by Eq. (3.102)

$$\frac{1}{Q} = \frac{1}{\sqrt{\omega\tau_\nu}} + \frac{\gamma - 1}{\sqrt{\omega\tau_\alpha}}$$

when $\omega\tau_\alpha \gg 1$ and $\omega\tau_\nu \gg 1$. In the present resonance tube, the theoretical value was given as 39, which again shows a good agreement with the experimental one.

The quality factor is also obtainable from the resonance curve. Figure 3.17 presents the relation between the square of end pressure $|p_e|^2$ and f. The pressure amplitude $|p_e|$ at the right end was measured as the driving frequency gradually changed, and then normalized by reference amplitude $p^* = 782$ Pa. Figure 3.17 also shows a result of the Lorenz curve fit. The quality factor turned out to be $Q = 25$ from the ratio of the resonance frequency and the width at the half maximum. This value was significantly lower than the actual one of $Q = 36$. Although the derivation of Q from the resonance curve assumes a constant velocity amplitude for a certain range of the frequency, this condition is not always satisfied because of a frequency characteristic of the driver, even when a constant voltage or current is maintained during the experiments. Careful experiment would be necessary to determine the quality factor from the resonance curve, although it provides a simple and easy method.

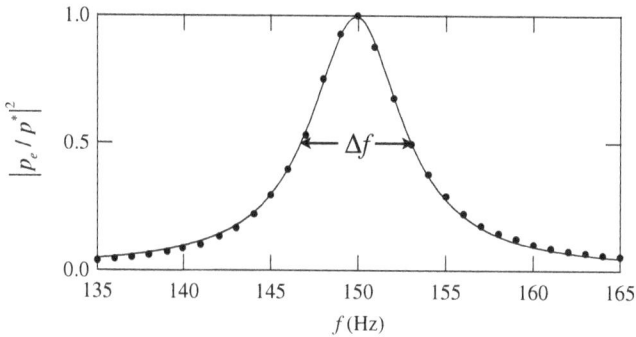

Figure 3.17: Experimental resonance curve. The curve represents the result of Lorentz curve fitting.

3.7 Acoustic Resonance Tube with Temperature Gradient

In a resonance tube with uniform temperature, oscillations of the gas column decay with time because of inevitable energy dissipations due to the viscosity and thermal conductivity of the gas. The gas column reaches the quiescent state in the end unless the resonance tube is sourced by an external power source like an acoustic driver. Oscillations of the gas column, however, can be maintained without the external driver, if the axial temperature gradient is present in the resonance tube. The appearance of thermoacoustic oscillations is considered here through the quality factor.

3.7.1 *Taconis oscillation*

If the resonance tube has a non-uniform temperature distribution along its axis, the gas column can spontaneously start to oscillate when the temperature difference between hot and cold parts goes beyond some critical value. One of such phenomena is Taconis oscillation, which refers to spontaneous gas oscillations in a transfer tube immersed into a liquid He vessel from the outside at room temperature. This phenomenon triggered a series of studies that tried to extend the acoustic theory to include the problem of acoustic gas

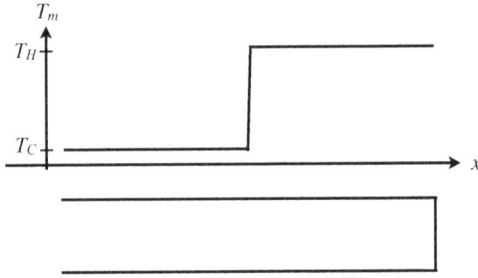

Figure 3.18: Temperature distribution of Kramers model.

oscillations in non-uniform temperature tubes. Kramers [52] assumed a step-function like temperature distribution on the tube wall as shown in Fig. 3.18, and adopted a wide tube approximation to make a stability analysis from hydrodynamic point of view. Rott used the same temperature distribution but obtained a reasonable theoretical stability curve of the oscillation [6] without relying on the wide tube approximation.

Rott's theoretical stability curve was experimentally verified by Yazaki of Tsukuba University [53]. Yazaki used a U-shaped tube with both ends closed in place of a tube with one end open and the other end closed. The mean pressure p_m in the closed tube was controlled to vary the parameter $\omega \tau_\alpha$ over a wide range, through the pressure dependence of thermal diffusivity α of the working gas of Helium. The U-shaped tube with both ends closed also made the problem free from the open-end correction. He made a great effort to obtain the step-function like temperature distribution. Experiments were conducted as follows: under a constant temperature ratio, the value of $\omega \tau_\alpha$ was gradually changed through the mean pressure p_m, which decreased the pressure amplitude of the unstable gas oscillation. The relation between T_H/T_C and $\omega \tau_\alpha$ when the quiescent state became stable determined the experimental stability curve.

Figure 3.19 shows the results obtained with various positions of temperature discontinuity. The quiescent state of the system is unstable inside of the curve and is stable outside of it. A good agreement between the experimental data and the theoretical prediction lends

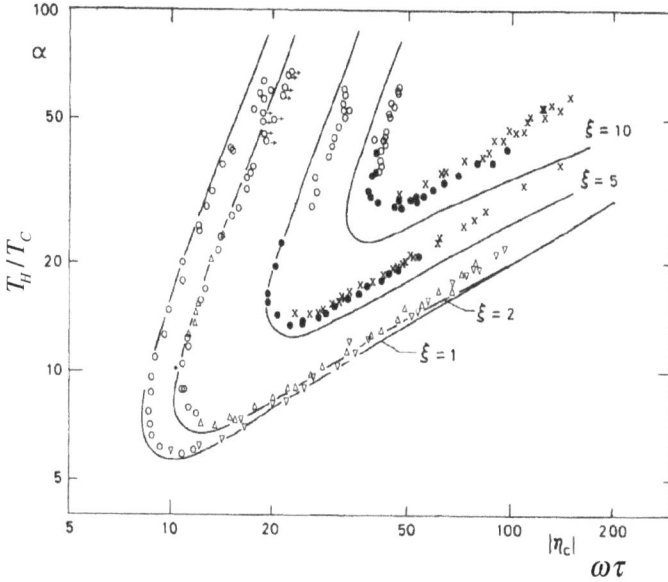

Figure 3.19: Stability curve of Taconis oscillation [53].

strong support to Rott's theory on thermoacoustic spontaneous gas oscillation.

3.7.2 Quality factor of resonance tube with temperature gradient

How does the quality factor Q change when the system approaches the stability limit by the increase of the temperature difference? To answer this question, Atchley [54, 55] studied the resonance curve from the pressure p_e measured at the closed end of the driven resonance tube when the axial temperature difference was imposed locally on the tube. He found that the inverse of the quality factor, $1/Q$, decreased linearly with increasing ΔT, and that it became zero when ΔT went to the critical value ΔT^* at the onset of the oscillations, as shown in Fig. 3.20. This result indicates that Q increases infinitely large at ΔT^*. In terms of the relaxation time $\tau_E = Q/\omega_0$, it can be said that acoustic fluctuations never stop at ΔT^*, because $\tau_E \to \infty$.

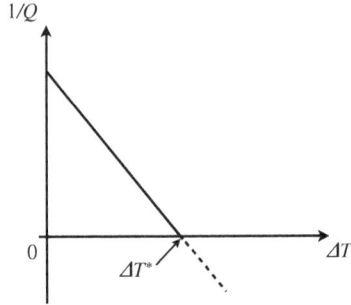

Figure 3.20: Relation between $1/Q$ and ΔT.

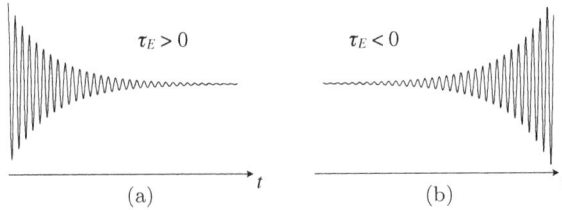

Figure 3.21: Temporal variations of (a) damped oscillations and (b) self-sustained oscillations.

Atchley also attained the temporal variation of p_e when $\Delta T > \Delta T^*$ [56]. The oscillations initially increased exponentially with time as shown in Fig. 3.21(b), in contrast to the damped oscillations with $\Delta T < \Delta T^*$ in (a). This result indicates that τ_E changes its sign from positive to negative as ΔT goes beyond ΔT^*, and also that $1/Q$ becomes negative as shown by a dashed line in Fig. 3.20.

From an energetic point of view, the temperature dependence of $1/Q$ in Fig. 3.20 can be interpreted through Eq. (3.1)

$$\frac{1}{Q} = \frac{W}{\omega E}.$$

The decrease of $1/Q$ means that the power W externally supplied to the resonance tube decreases with increasing ΔT when constant ω and E were maintained. When ΔT reaches ΔT^*, W goes to zero with non-zero E; the oscillations can be sustained without the external power supply. Finally with $\Delta T > \Delta T^*$, the resonance tube starts to produce extra power to amplify the oscillations by itself. In other words, the externally-heated resonance tube can serve as a power source, just like heat engines. In the next chapter, we consider this problem from thermodynamical point of view, to establish a concept of *acoustical heat engine* that produces sound from heat.

3.8 Appendix

3.8.1 *Analogy between acoustical system and electrical system*

The equation of motion for an inviscid fluid moving back and forth periodically with angular frequency ω [Eq. (3.51)] is written in a complex form as

$$u_1 = \frac{i}{\omega \rho_m} \frac{dp_1}{dx}.$$

If we consider a gas element occupying a short segment of length l, it can be expressed in a difference form as $\Delta p_1 = -i\omega \rho_m l u_1$, where $\Delta p_1 = p_1(x+l) - p_1(x)$ denotes the pressure difference at ends of the gas element; U_1 represents the complex amplitude of the volume velocity $U' = Au'$ given by using the cross-sectional area A of the tube. The equation of motion is then expressed as

$$\Delta p_1 = -i\omega L U_1$$

where

$$L = \frac{\rho_m l}{A}$$

is called inertance of the tube.

The equation of continuity is given by

$$\frac{1}{c_S^2} \frac{\partial p'}{\partial t} + \rho_m \frac{\partial u'}{\partial x} = 0.$$

By introducing complex representation for the acoustic variables, we have

$$i\omega p_1 + \rho_m c_S^2 \frac{du_1}{dx} = 0,$$

which is expressed in a difference form for a segment of length l as

$$p_1 = -\frac{\Delta U_1}{i\omega C}$$

using

$$C = \frac{Al}{\rho_m c_S^2} = AlK_S.$$

Here, ΔU_1 represents the difference of volume velocity between $U_1(x)$ and $U_1(x+l)$; $C = AlK_S$ is called compliance of the tube.

As shown in Fig. 3.22, voltage and current in an ac electrical circuit can be seen analogous to pressure and volume velocity, respectively. Then, the inertance of the tube plays the role of an inductance while the compliance plays the role of a capacitance, respectively. Such analogy can transform the problem of acoustic wave propagation to the problem of electrical circuits, which often makes it easy to handle the problem.

As we have seen earlier, a tube segment in acoustic systems plays both roles of inductance and capacitance in electrical circuits. Which role, inertance or compliance, does dominate in a given acoustic system? Let's consider this problem from an energetic point of view.

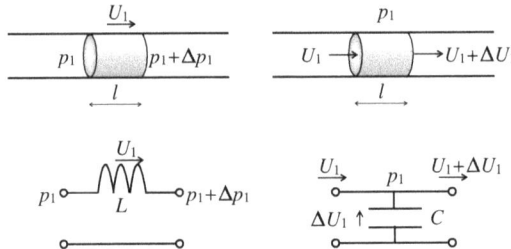

Figure 3.22: Schematic illustration of acoustical system and electrical system.

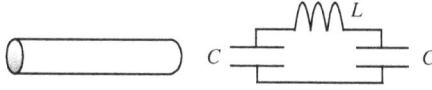

Figure 3.23: Acoustic resonance tube with both ends closed and the analogous circuit.

The ratio of the time-averaged potential energy over the time-averaged kinetic energy is given from Eq. (3.43) as

$$\frac{\frac{1}{2}K_S\langle p'^2\rangle_t}{\frac{1}{2}\rho_m\langle u'^2\rangle_t} = \frac{K_S}{\rho_m}\left|\frac{p_1}{u_1}\right|^2 = \left|\frac{z}{\rho_m c_S}\right|^2$$

where $z = p_1/u_1$ means the complex specific acoustic impedance of the tube element. From this ratio, we can say as follows: If the magnitude $|z|$ is much greater than the characteristic impedance $\rho_m c_S$, the tube segment can be seen as a compliance, whereas if $|z|$ is much smaller than $\rho_m c_S$, it can be seen as an inertance.

As an example of application of the analogy, consider an acoustic resonance tube of length l_0 and cross-sectional area A, whose ends are closed by rigid plates, as shown in Fig. 3.23. One can imagine that the pressure maximum is created at the end, and the velocity maximum is in the middle of the resonance tube in the fundamental oscillation mode. Therefore, the analogous electrical circuit would be constructed from two capacitances of C at ends and an inductance L in the middle. The natural frequency f_0 is given from the equation

$$i\omega L + \frac{2}{i\omega C} = 0, \quad \text{with} \quad L = \frac{\rho_m l_0}{2A}, \quad C = \frac{K_S A l}{4}.$$

Such a solution implies $f_0 = 2c/(\pi l_0)$, which is 1.27 times larger than the exact value of $f_0 = c/(2l_0)$, but still serves as a reasonable estimation.

3.8.2 *Acoustic field with boundary conditions*

We introduce here a derivation of acoustic field when boundary conditions are given in a more general way than Section 3.4.1,

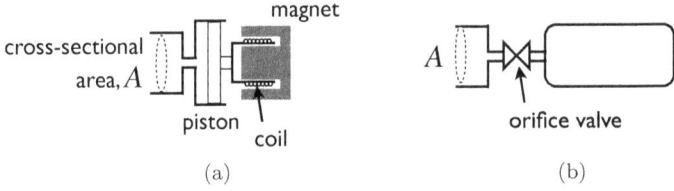

Figure 3.24: Acoustic resonance tube terminated by piston with coil (a) and by orifice with gas tank (b).

where the resonance tube was closed simply by a rigid plate. The thermoacoustic devices are often terminated by a linear alternator made of solid piston and a voicecoil to extract electricity from the acoustic engine, or by an orifice valve with a buffer tank as in the case of a pulse tube refrigerator, as shown in Fig. 3.24. These end elements lead to non-zero velocity at the end of the tube, in contrast to the rigid plate termination. So the problem here is how to determine the acoustic field when the tube end is closed by an arbitral acoustic impedance that is given by $z_e = Ap_e/V_e$, where p_e denotes the end pressure, $V_e = Au_{1r}$ means the volume velocity flowing into the end elements like the piston and the orifice with the tank, and A is the tube cross-sectional area. Note that z_e is a complex number when the end elements are characterized with acoustical inertance and/or compliance.

The acoustic field in the resonance tube is determined as follows. The axial coordinate x is directed to the right, and the left end is taken as $x = 0$. The resonance tube is connected to z_e at the right end $x = L$, as shown in Fig. 3.25. Suppose that the pressure oscillation $p'(x,t) = \text{Re}[p_1(x)e^{i\omega t}]$ has the complex amplitude p_1 given by

$$p_1 = C_+ e^{-ik(x-L)} + C_- e^{ik(x-L)}$$

where C_+ and C_- are the unknown constants, and k is the wavenumber [Eq. (3.88)] given by

$$\frac{k}{k_S} = \sqrt{\frac{1 + (\gamma - 1)\chi_\alpha}{1 - \chi_\nu}}.$$

The complex amplitude u_{1r} of the cross-sectional mean velocity is expressed from the equation of motion [Eq. (3.80)] as

$$u_{1r}(x) = \frac{i(1 - \chi_\nu)}{\omega \rho_m} \left[-ikC_+ e^{-ik(x-L)} + ikC_- e^{ik(x-L)} \right]$$

$$= \frac{C_+ e^{-ik(x-L)} - C_- e^{ik(x-L)}}{z}$$

where z denotes the specific acoustic impedance given by

$$z = \frac{\omega \rho_m}{k(1 - \chi_\nu)}.$$

If the end pressure $p_1(L)$ is set to $p_1(L) = p_e$, we have

$$p_e = C_+ + C_-.$$

The velocity $u_{1r}(L)$ at the end is expressed as

$$u_{1r}(L) = \frac{C_+ - C_-}{z}.$$

The specific acoustic impedance z_e at the right end is written as

$$z_e = \frac{p_e z}{C_+ - C_-}.$$

So the coupled equations with regard to C_+ and C_- are derived as

$$C_+ + C_- = p_e$$

$$C_+ - C_- = p_e \frac{z}{z_e}.$$

Solving these yields

$$C_+ = \frac{1 + \Gamma}{2} p_e, \quad C_- = \frac{1 - \Gamma}{2} p_e$$

with

$$\Gamma = \frac{z}{z_e}.$$

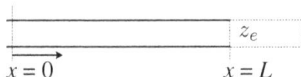

Figure 3.25: Acoustic resonance tube terminated with the component with z_e.

When the right end is terminated by a rigid plate, z_e can be assumed as $1/z_e = 0$, which gives $\Gamma = 0$. In this case, we have

$$C_+ = C_- = \frac{1}{2}.$$

When the end is closed by an acoustic terminator with $z_e = z$, we have $\Gamma = 1$. Thus,

$$C_+ = 1, \quad C_- = 0,$$

which corresponds to a forward traveling wave without the reflected wave. Various end components provide different acoustic impedance z_e at the end of the tube. In Chapter 8, we present examples of the terminating impedance to promote the performance of pulse tube refrigerators.

The acoustic field can also be determined if pressure and velocity at a point x is given. For simplicity, we assume that the complex amplitudes of pressure and velocity are given by p_0 and v_0 at $x = 0$. By setting the pressure as

$$p_1(x) = C_+ e^{-ikx} + C_- e^{ikx},$$

the equation of motion gives

$$u_{1r}(x) = \frac{C_+ e^{-ikx} - C_- e^{ikx}}{z}.$$

The constants C_+ and C_- are determined by the boundary conditions $p_1(0) = p_0$ and $u_{1r}(0) = v_0$:

$$C_+ + C_- = p_0, \quad C_+ - C_- = zv_0.$$

From these relations, we obtain

$$C_+ = (p_0 + zv_0)/2, \quad C_- = (p_0 - zv_0)/2.$$

Therefore, the pressure is expressed as

$$p_1(x) = \frac{p_0 + zv_0}{2}e^{-ikx} + \frac{p_0 - zv_0}{2}e^{ikx}$$

$$= \frac{e^{-ikx} + e^{ikx}}{2}p_0 + \frac{e^{-ikx} - e^{ikx}}{2}zv_0.$$

Namely,

$$p_1(x) = (\cos kx)p_0 - iz(\sin kx)v_0.$$

The cross-sectional mean velocity is also expressed as

$$u_{1r}(x) = \frac{1}{z}\left(\frac{p_0 + zv_0}{2}e^{-ikx} - \frac{p_0 - zv_0}{2}e^{ikx}\right)$$

$$= \frac{e^{-ikx} - e^{ikx}}{2z}p_0 + \frac{e^{-ikx} + e^{ikx}}{2z}v_0.$$

Namely,

$$u_{1r}(x) = \frac{\sin kx}{iz}p_0 + (\cos kx)v_0.$$

In a matrix form, $p(x)$ and u_{1r} are written as

$$\begin{pmatrix} p(x) \\ u_{1r}(x) \end{pmatrix} = M \begin{pmatrix} p_0 \\ v_0 \end{pmatrix},$$

where M denotes the transfer matrix given by

$$M = \begin{pmatrix} \cos kx & -iz\sin kx \\ \dfrac{\sin kx}{iz} & \cos kx \end{pmatrix}.$$

As we explain in Chapter 9, the transfer matrix provides a useful tool to examine the acoustic field in the thermoacoustic system consisting of various types of tubes like the resonance tube, heat exchanger, and regenerator. For this particular purpose, the volume velocity $U = Au_{1r}$ is employed in place of the cross-sectional mean velocity u_{1r}, because the volume velocity is assumed to be conserved at the junction of two tube segments. The relations between

$(p(x), U(x))$ and $(p(x + \Delta x), U(x + \Delta x))$ is then expressed as

$$\begin{pmatrix} p(x + \Delta x) \\ U(x + \Delta x) \end{pmatrix} = M \begin{pmatrix} p(x) \\ U(x) \end{pmatrix},$$

where M is given by

$$M = \begin{pmatrix} \cos k\Delta x & \dfrac{-iz \sin k\Delta x}{A} \\ A\dfrac{\sin k\Delta x}{iz} & \cos k\Delta x \end{pmatrix}$$

for a tube segment of length Δx and uniform temperature.

3.8.3 Oscillatory flow velocity over a plate

We investigate here the viscous flow over plate, as shown in Fig. 3.26. The equation of motion in x direction is given by

$$\rho_m \frac{\partial u'}{\partial t} = -\frac{\partial p'}{\partial x} + \mu \frac{\partial^2 u'}{\partial z^2}.$$

The non-slip condition at the plate surface provides the boundary condition:

$$u' = 0 \quad \text{for } z = 0.$$

By assuming harmonic oscillations with an angular frequency ω, the equation of motion is rewritten as

$$u_1 + \frac{i\nu}{\omega} \frac{d^2 u_1}{dz^2} = \frac{i}{\omega \rho_m} \frac{dp_1}{dx},$$

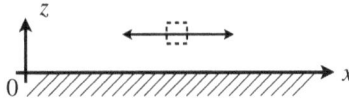

Figure 3.26: Oscillating gas particle over a plate.

where u_1 and p_1 represent the complex amplitude of velocity and pressure, respectively. The general solution of the homogeneous differential equation when the right hand side is set to zero is

$$u_1 = C_1 e^{-\lambda z} + C_2 e^{\lambda z},$$

where $\lambda = (1+i)/\delta_\nu$ and $\delta_\nu = \sqrt{2\nu/\omega}$. In order for u_1 to remain finite as z goes to infinitely large, the constant C_2 must be zero. Therefore,

$$u_1 = C_1 e^{-\lambda z} + \frac{i}{\omega \rho_m} \frac{dp_1}{dx}.$$

The constant C_1 is determined by the boundary condition. Finally the solution is given by

$$u_1 = \frac{i}{\omega \rho_m} \frac{dp_1}{dx} \left(1 - e^{-\lambda z}\right) = \frac{i}{\omega \rho_m} \frac{dp_1}{dx} \left\{ 1 - \exp\left[-(1+i)\frac{z}{\delta_\nu}\right]\right\}.$$

Figure 3.27 shows the velocity distribution along the z axis, where Ω is taken as $\Omega = 1 - \exp[-(1+i)z/\delta_\nu]$. The magnitude $|\Omega|$ rapidly increases near $z/\delta_\nu \approx 1$ and then shows a maximum at $z/\delta_\nu \approx 2.3$. Finally, it reaches almost unity in the region with $z/\delta_\nu > 5$. Therefore, δ/ν means the viscous boundary layer thickness. For comparison, in the case of the oscillating flow in a circular tube, the core region was created in the range $r/\delta_\nu > 3$ from the tube wall when r_0/δ_ν is sufficiently high.

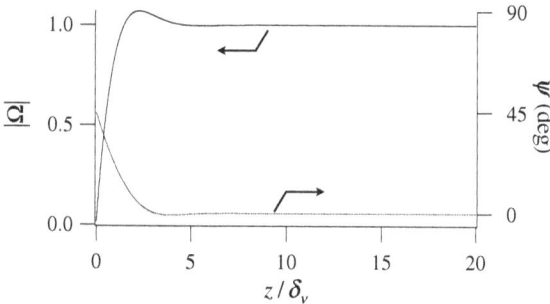

Figure 3.27: Velocity profile of gas particle oscillating over a plate. Ψ indicates the phase of Ω; $\Psi = \arg \Omega$.

3.9 Problems

1 Consider a damped oscillation $u(t) = Ae^{-t/(2\tau)}\cos(\omega_0 t)$, when $t \geq 0$, otherwise $u(t) = 0$, where A, τ and ω_0 are constants.

(1) Derive the Fourier transform $U(\omega)$ of $u(t)$, where

$$U(\omega) = \int_{-\infty}^{\infty} x(t)e^{-i\omega t}\,dt.$$

(2) When $U(\omega)$ is approximated by

$$U(\omega) = \frac{A/2}{1/(2\tau) + i(\omega - \omega_0)},$$

show that

$$|U(\omega)|^2 = \frac{A^2}{\omega_0^2}\frac{1}{4\epsilon^2 + 1/(\omega_0 \tau)^2},$$

where $\epsilon = \dfrac{\omega - \omega_0}{\omega_0}$.

2 (Two-sensor method) Consider a mono-frequency plane pressure wave $p(x,t) = \mathrm{Re}[p_1(x)e^{i\omega t}]$ in a tube filled with a gas. Suppose that acoustic pressures $P_A(t) = \mathrm{Re}[p_A e^{i\omega t}]$ and $P_B(t) = \mathrm{Re}[p_B e^{i\omega t}]$ are measured simultaneously at positons A and B separated by axial distance Δx. When the complex amplitude is expressed by $p_1 = C_+e^{-ikx} + C_-e^{ikx}$, determine the constants C_+ and C_- using the complex amplitudes p_A and p_B, where k denotes the wavenumber.

Answers

1 (1) Since $u(t)$ is transformed as $u(t) = \dfrac{A}{2}e^{-t/(2\tau)}\left(e^{i\omega_0 t} + e^{-i\omega_0 t}\right)$, we obtain

$$U(\omega) = \frac{A}{2}\int_0^{\infty} e^{\{-1/(2\tau)-(\omega-\omega_0)\}t} + e^{\{-1/(2\tau)-(\omega+\omega_0)\}t}\,dt$$

$$= \frac{A}{2}\left\{\frac{1}{1/(2\tau) + i(\omega - \omega_0)} + \frac{1}{1/(2\tau) + i(\omega + \omega_0)}\right\}.$$

(2) As $X(\omega)$ becomes large only in the vicinity of $\pm\omega_0$, it is approximated by

$$X(\omega) = \frac{A/2}{1/(2\tau) + i(\omega - \omega_0)},$$

when $\omega \approx \omega_0$. Therefore,

$$|X(\omega)|^2 = \left(\frac{A}{2}\right)^2 \frac{1}{1/(4\tau^2) + (\omega - \omega_0)^2}.$$

By introducing $\epsilon = \dfrac{\omega - \omega_0}{\omega_0}$, we obtain

$$|X(\omega)|^2 = \frac{A^2}{\omega_0^2} \frac{1}{4\epsilon^2 + 1/(\omega_0\tau)^2}.$$

This result means that the square of the amplitude spectrum of a damped oscillation is equivalent to the resonance curve of a forced oscillation, and therefore, the quality factor Q obtained from the relaxation time of the damped oscillation and that from the response curve of the forced oscillation are the same. (The resonance curve of a mass-damper-spring system is given in Section 3.2.)

2 The complex amplitudes p_A and p_B are respectively given by

$$p_A = p_1\left(-\frac{\Delta x}{2}\right) = C_+ e^{ik\Delta x/2} + C_- e^{-ik\Delta x/2}$$

and

$$p_B = p_1\left(\frac{\Delta x}{2}\right) = C_+ e^{-ik\Delta x/2} + C_- e^{ik\Delta x/2}.$$

By eliminating C_-, we obtain

$$C_+ = \frac{p_A e^{i\phi} - p_B e^{-i\phi}}{2\sinh(2i\phi)}$$

where $\phi = \dfrac{k\Delta x}{2}$. In the same way, we have

$$C_- = -\frac{p_A e^{-i\phi} - p_B e^{i\phi}}{2\sinh(2i\phi)}.$$

In this way, the acoustic pressure

$$p_1(x) = C_+ e^{-ikx} + C_- e^{ikx}$$

can be derived from the measured pressures p_A and p_B. Once $p_1(x)$ is determined, it is easy to obtain $u_1(x)$. See references [20, 21].

Chapter 4

From Acoustics to Thermoacoustics

Heat flow and work flow are central concepts of thermoacoustics. They are derived by taking the time average of the energy equation of hydrodynamics for a periodically oscillating fluid. The space integral of the time-averaged energy equation provides a link between heat flow and work flow with the first law of thermodynamics. Heat flow and work flow provide a basis for discussing thermoacoustic devices from the viewpoint of a heat engine, and also contribute to renewal of a conventional conceptual diagram of heat engines.

4.1 Ceperley's Proposal

A Stirling engine with an alpha type configuration consists of a pair of opposed pistons, a regenerator, and cold and hot heat exchangers, as shown in Fig. 4.1(a). The engine is filled with air, helium, or hydrogen gas as the working gas. The pistons on the cold side and on the hot side are often called a compression piston and an expansion piston, respectively. The work done by the piston on the cold side is compression work, whereas that by the piston on the hot side is expansion work.

The two pistons are mechanically connected to each other so that they move back and forth with the same frequency but with a phase angle of about 90°. The resulting motion of the pistons drive the working gas to execute a Stirling thermodynamic cycle, where the gas undergoes a series of compression/expansion and heating/cooling processes. If the gas flow channels in the regenerator are sufficiently small to warrant isothermal processes and if the viscous losses are

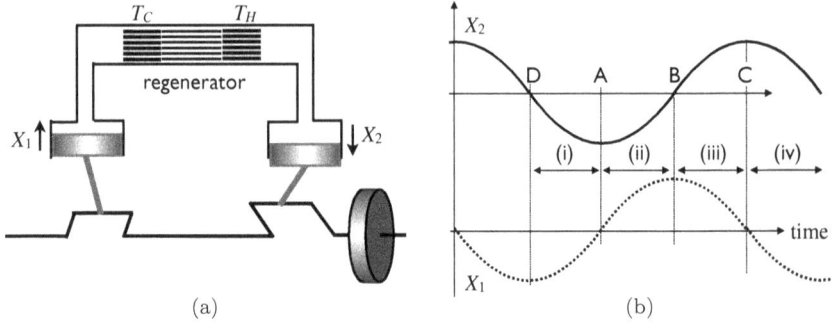

(a)   (b)

Figure 4.1: Schematic illustration of (a) the Stirling engine and (b) temporal motions X_1 and X_2 of the two pistons. The working gas is confined between the pistons, and therefore, it undergoes thermodynamic processes in order of (i) compression, (ii) heating due to the displacement to the hot side, (iii) expansion, and (iv) cooling due to the displacement to the cold side. Both the gas pressure and displacement oscillate in time, keeping a phase difference of 90°. The velocity oscillation leads the displacement by 90°, and hence the pressure and velocity oscillations are in phase with each other.

negligibly small, the cycle can be seen as a reversible thermodynamic cycle. Owing to the execution of the cycle, the expansion work becomes larger than the compression work by an amount of the output work, which can be extracted from the engine as the shaft work.

During the operation of the Stirling engine, the pressure and velocity of the working gas oscillate essentially in phase with each other, in the same way as those in acoustic traveling waves of the gas. Ceperley noticed this phase relationship and proposed a *pistonless Stirling engine* where oscillating mechanical pistons were replaced with the acoustic traveling waves [31].

It would be intuitively possible to see some similarity between the gas motions in the Stirling cycle and in the acoustic waves, because both of them are associated with the periodic oscillations of pressure and velocity. Indeed, Rayleigh gave a qualitative explanation on the Rijke tube in the following way:

> "at the phase of greatest condensation heat is received by the air, and at the phase of greatest rarefaction heat is given up from it, and thus there is a tendency to maintain the vibrations."

It was, however, Ceperley that gave a quantitative theoretical discussion on this topic for the first time. Although he failed to experimentally verify the concept of pistonless Stirling engine, his thermodynamic approach stimulated Wheatley in Los Alamos laboratory. His group reexamined the thermoacoustic theory presented by Rott [6–12] from a thermodynamic point of view [15]. Tominaga of Tsukuba University also proposed the thermoacoustic theory [16] individually. The present thermoacoustic theory greatly owes its success to their achievements. Before going into details of the thermoacoustic theory, let's review the thermodynamic description of heat engines.

4.2　Conventional Heat Engine Concept

4.2.1　*The first and second law of thermodynamics*

Two heat baths with different temperatures are necessary for the operation of a heat engine. The heat bath means an ideal heat source that can supply or receive any amount of heat without changing its temperature. There are two categories in heat engines: one is a prime mover, and the other is a heat pump. The prime mover receives the input heat Q_H from the heat bath of temperature T_H, and converts a part of it to output work W, while delivering the rest heat Q_C to the heat bath of temperature T_C ($T_C < T_H$). On the other hand, the heat pump receives input work W to absorb heat Q_C from the heat bath of T_C and transfer heat Q_H to the heat bath of T_H. Figure 4.2 shows schematic diagrams for the prime mover and the heat pump.

The first law of thermodynamics states the energy conservation law expressed in terms of heat and work. More specifically, it is written for both the prime mover and the heat pump as

$$Q_H = Q_C + W. \tag{4.1}$$

When heat Q_H enters the prime mover from the heat bath of temperature T_H, the entropy of the heat bath decreases by the amount of $S_H = Q_H/T_H$. If this process is reversible, the entropy of the prime mover increases by the same amount. On the other

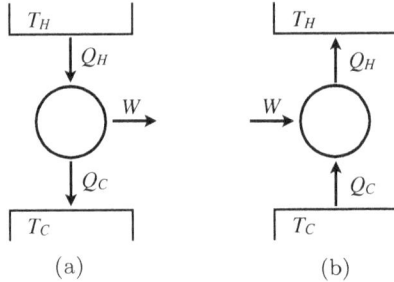

Figure 4.2: Schematic diagrams of a (a) prime mover and (b) heat pump. The center circles symbolize the thermodynamic cycles.

hand, when heat Q_C goes out of the prime mover into the cold heat bath of temperature T_C reversibly, the entropy of the prime mover decreases by the amount of $S_C = Q_C/T_C$. For a steadily operating reversible heat engine, $S_H = S_C$ is satisfied. For an actual heat engine, however, the entropy of the prime mover must increase because of the unavoidable irreversible processes and hence the leaving entropy must be greater than the entering entropy. The second law of thermodynamics states that

$$\Delta S = S_C - S_H \geq 0 \qquad (4.2)$$

for the prime mover, and

$$\Delta S = S_H - S_C \geq 0 \qquad (4.3)$$

for the heat pump. Here, ΔS signifies the entropy production of the heat engine. The relation $S_H = S_C$ holds only for a reversible heat engine. Otherwise, ΔS should be always positive.

The second law of thermodynamics gives the upper limit of the efficiency of heat engines. The efficiency of the prime mover is defined by output work W divided by input heat Q_H as

$$\eta = \frac{W}{Q_H}. \qquad (4.4)$$

Inserting the relation $W = Q_H - Q_C$ of the first law of thermodynamics yields

$$\eta = 1 - \frac{Q_C}{Q_H}. \tag{4.5}$$

From the second law of thermodynamics, the relation $Q_C/Q_H \geq T_C/T_H$ must hold. Hence, the efficiency should obey the following relation:

$$\eta \leq \eta_{\text{Carnot}} = 1 - \frac{T_C}{T_H}. \tag{4.6}$$

The *Carnot efficiency*, η_{Carnot}, gives the upper limit of the efficiency of the prime mover. Therefore, we often use the ratio of the efficiency to the Carnot efficiency, which is called the *second law efficiency*.

In the case of the heat pump, the efficiency is called COP (coefficient of performance). When the heat pump operates as a cooler that pumps up heat Q_C from the cold heat bath to the hot one, COP is given by

$$\text{COP} = \frac{Q_C}{W}. \tag{4.7}$$

From the first and the second laws of thermodynamics, we can see the upper limit of the COP is given by

$$\text{COP}_{\text{Carnot}} = \frac{T_c}{T_H - T_C}. \tag{4.8}$$

Therefore, actual coolers must satisfy the relation

$$\text{COP} \leq \text{COP}_{\text{Carnot}}, \tag{4.9}$$

meaning that the COP of any cooler cannot exceed $\text{COP}_{\text{Carnot}}$.

4.2.2 *Thermodynamic cycle*

The energy conversion between heat and work is sustained by the thermodynamic cycles that the working gas executes. The Carnot cycle is the representative one. Figure 4.3(a) shows the pressure p versus the volume V diagram for a gas confined in a cylinder plugged with a movable piston. The gas is adiabatically compressed

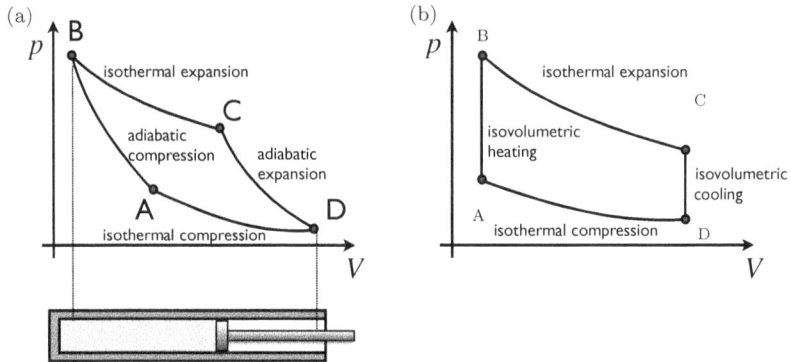

Figure 4.3: p–V diagrams of (a) Carnot cycle and (b) Stirling cycle. During the processes of A–B and C–D, the working gas experiences heating and cooling as it moves along the differentially heated regenerator while keeping thermal contacts with it. These processes are approximated by isovolumetric ones with reference to [57].

in the process of A–B, and isothermally heated by the heat bath of temperature T_H in the process of B–C. In the next process of C–D, the gas is adiabatically expanded and finally in the process of D–A, the gas is isothermally cooled by the heat bath of temperature T_C. All of these processes are thermodynamically reversible processes that are free from entropy production, and therefore, the Carnot efficiency should be attained.

If the adiabatic compression and expansion processes are replaced with the isovolumetric (isochoric) compression and expansion processes, respectively, the resulting cycle is called the Stirling thermodynamic cycle, which is illustrated in Fig. 4.3(b). Since only the reversible processes are involved between the two heat baths, the Stirling cycle can also achieve Carnot efficiency.

In contrast to reversible engines, irreversible processes are inevitably accompanied in the cycle in actual heat engines. The efficiency is lower than Carnot efficiency because of entropy production. Regardless of whether the cycle is reversible or irreversible, all the state quantities like pressure and volume, as well as entropy, goes back to the original value after a cycle. If we focus on the internal energy of the working fluid, its change

$$dU = TdS - pdV \qquad (4.10)$$

is zero after a cycle. Namely,

$$\oint dU = 0 \qquad (4.11)$$

holds. In other words, we have the relation

$$\oint T dS = \oint p dV, \qquad (4.12)$$

which represents that the net heat absorbed by the working fluid is equal to the net work done by the working fluid. Thus, Eq. (4.12) represents the mutual conversion between heat and work. The output work over one cycle is given by

$$W = \oint p dV. \qquad (4.13)$$

4.3 Thermoacoustic Representation of Heat Engines

The basic equations of thermodynamics as well as the first and second laws of thermodynamics treat the heat engine as a kind of a black box. Involved quantities of Q_H, Q_C, and W, and also T_H and T_C, are defined at the interface of the heat engine and the outer system, as in Fig. 4.2. The state quantities like pressure and temperature in the preceding section had no spatial variables. Therefore, any information inside the heat engine is not discussed in the framework of thermodynamics; it describes only a *macroscopic* picture of the heat engine, which may be insufficient to account for thermoacoustic phenomena.

For example, let's consider the self-sustained gas oscillations in a thermoacoustic system like Taconis oscillation that we have seen in Chapter 3. When the gas column starts to oscillate spontaneously, one may think that the engine has started. But is it a correct statement from the thermodynamic point of view? If the gas is confined in a closed resonance tube or in a looped tube, the entire gas volume is always constant. Therefore, the right hand side of Eq. (4.12) is zero and hence no energy conversion should occur. The pressure must be uniform when one draws a picture like Fig. 4.3, but the pressure in spontaneous gas oscillations varies with time and space. In this way, it would be difficult to discuss the system of spontaneous gas oscillations as an acoustic heat engine.

Rott, who succeeded in describing Taconis oscillations quantitatively, discussed the acoustic variables of temperature, pressure, and velocity on the basis of the hydrodynamic equations. His theory provided a detailed understanding on the oscillatory dynamics of gas under viscous and thermal interactions with the wall, and succeeded in predicting the stability curve of Taconis oscillation. Merkli and Thomann used his theory to explain the cooling effect by acoustic gas oscillations observed in a resonance tube. Taconis oscillation is undoubtedly a prototype of an acoustic prime mover, whereas the experiments of Merkli–Thomann provide the prototype of an acoustic cooler. However, it is hard to imagine a picture of the heat engine from his theory, probably because hydrodynamics does not discuss the mutual conversion of heat and work, although it considers the total energy flux called enthalpy flow.

Wheatley and Tominaga proposed the concept of work flow $\tilde{I}$ and heat flow $\tilde{Q}$ to link the macroscopic picture of the heat engine in thermodynamics with the *microscopic* picture of gas parcels in hydrodynamics. As we see later, $\tilde{I}$ and $\tilde{Q}$ are formulated by taking time average of the hydrodynamic energy equation. Also by taking spatial average of $\tilde{I}$ and $\tilde{Q}$ over the cross-section and integrating them with respect to the axial position, we arrive at the expressions equivalent to the first law of thermodynamics. Thus, thermoacoustic phenomena can be seen as a result of mutual energy conversion between $\tilde{I}$ and $\tilde{Q}$: the acoustic prime mover executes the energy conversion from $\tilde{Q}$ to $\tilde{I}$, whereas the acoustic heat pump promotes that from $\tilde{I}$ to $\tilde{Q}$.

Figure 4.4 shows a hierarchical structure relevant to thermodynamics, fluid dynamics, and thermoacoustics. The first hierarchy is described by thermodynamics, where the macroscopic view is provided by the first and the second law of thermodynamics. The third hierarchy is comprised of the basic equations of hydrodynamics, where the microscopic view is presented for the dynamics of gas parcels. The second hierarchy is placed between the first and third ones, where the concepts of work flow and heat flow are used to

Third layer: fluid dynamics

$$\frac{\partial}{\partial t}\left(\rho\varepsilon + \frac{1}{2}\rho u^2\right) = -\nabla \cdot \left[\left(\rho h + \frac{1}{2}\rho u^2\right)u - \kappa\nabla T - u \cdot \Sigma\right]$$

$\cdots$

$p(x,t), u(x,t), T(x,t), S(x,t)$

Second layer: thermoacoustics

$$\frac{d}{dx}(\tilde{I} + \tilde{Q}) = 0$$

$$\frac{d\tilde{s}}{dx} \geq 0$$

$$w = \frac{d\tilde{I}}{dx}$$

$$\tilde{I}(x), \tilde{Q}(x), \tilde{H}(X) = \tilde{I}(x) + \tilde{Q}(x), \tilde{s}(x) = \tilde{Q}(x)/T$$

Fist layer: thermodynamics

$$Q_H = Q_C + W$$

$$\Delta S \geq 0$$

$$\oint T\,dS = \oint p\,dV \ (\equiv W)$$

$$Q_H, Q_C, T_H, T_C$$

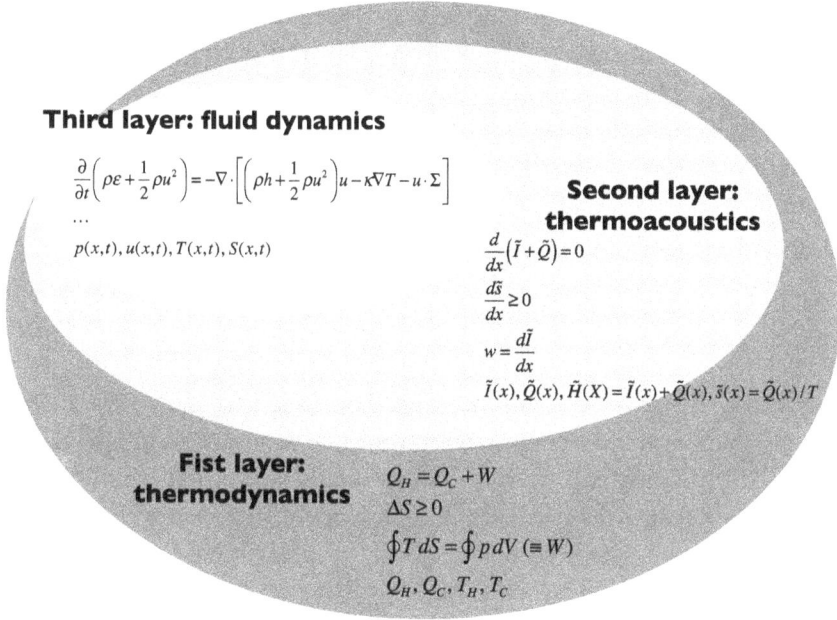

Figure 4.4: Hierarchical structure of thermoacoustics.

describe various thermoacoustic phenomena. That is to say, work flow and heat flow play the role of a ladder to link hydrodynamics and thermodynamics, from which one can formulate the energy conversions of acoustic heat engines by using the gas motions in time and space.

4.4 Energy Flux Density in a Periodically Steady Flow

4.4.1 *Enthalpy flux density, work flux density, and heat flux density*

The energy equation is expressed as

$$\frac{\partial}{\partial t}\left(\rho\epsilon + \frac{1}{2}\rho|u|^2\right) + \nabla \cdot \left[-\kappa\nabla T - u \cdot \Sigma + \left(\rho h + \frac{1}{2}\rho|u|^2\right)u\right] = 0.$$

$$(4.14)$$

The first term on the left hand side represents the time derivative of energy density, and the second term gives the divergence of energy flux density. Symbols ϵ and h represent internal energy and enthalpy of the fluid per unit mass, and $\boldsymbol{\Sigma}$ means the viscous stress tensor.[1]

In a periodically steady flow, all the state quantities should repeat themselves after a period of oscillation. Thus, taking the time average of Eq. (4.14) yields

$$\nabla \cdot \left\langle -\kappa \nabla T - \boldsymbol{u} \cdot \boldsymbol{\Sigma} + \left(\rho h + \frac{1}{2}\rho|\boldsymbol{u}|^2 \right) \boldsymbol{u} \right\rangle_t = 0, \qquad (4.15)$$

where $\langle \ \rangle_t$ denotes the time average. For relatively slow flow fields, the terms including viscous stress tensor and kinetic energy are safely assumed to be negligibly small. Also, when the thermal conductivity of the gas is taken as a constant independently of temperature, the term $\langle -\kappa \nabla T \rangle_t$ is replaced with $-\kappa \nabla T_m$, which represents the energy flux density due to the mean temperature T_m. Therefore, the time-averaged energy flux density associated with the oscillatory motion of the fluid is the enthalpy flux density given by

$$\langle \rho h \boldsymbol{u} \rangle_t = \langle (\rho_m + \rho')(h_m + h')\boldsymbol{u}' \rangle_t$$

where h_m represents the time average of the enthalpy of the fluid, and h' represents the fluctuation around it. The enthalpy flux density is divisible into the axial component and the radial component, but we are concerned only with the axial component because we consider the gas flowing in a rigid-walled tube.

The axial component of time-averaged enthalpy flux density is expressed as $\langle (\rho_m + \rho')(h_m + h')u' \rangle_t$ by using the axial velocity u' of

---

[1]When the velocity is expressed as $\boldsymbol{u} = (u, v, w)$ in Cartesian coordinates, the viscous stress tensor $\boldsymbol{\Sigma}$ is written by using viscosity μ and bulk viscosity ζ as

$$\Sigma_{xx} = \mu \left[2\frac{\partial u}{\partial x} - \frac{2}{3}(\nabla \cdot \boldsymbol{u}) \right] + \zeta(\nabla \cdot \boldsymbol{u})$$

$$\Sigma_{xy} = \mu \left[\frac{\partial u}{\partial y} + \frac{\partial v}{\partial x} \right].$$

the velocity u'. The cross-sectional average is then given by

$$\langle\langle(\rho_m + \rho')(h_m + h')u'\rangle\rangle$$

where the cross-sectional average of a certain quantity $X = X(t, r)$ for a circular tube with radius r_0 is given by

$$\langle X\rangle_r = \frac{1}{\pi r_0^2}\int_0^{r_0} 2\pi r X \, dr.$$

It should be noted that the relation $\langle\langle X\rangle\rangle = \langle\langle X\rangle_r\rangle_t = \langle\langle X\rangle_t\rangle_r$ holds because t and r are independent of each other. The axial enthalpy flux density averaged over the cross-section and time is transformed as

$$\langle\langle(\rho_m + \rho')(h_m + h')u'\rangle\rangle = h_m\langle\langle(\rho_m + \rho')u'\rangle\rangle + \rho_m\langle\langle h'u'\rangle\rangle + \langle\langle\rho'h'u'\rangle\rangle.$$

The first term on the right hand side includes the mass flux density averaged over the cross-section and time, and it should be zero for a periodically steady flow without a total mass flow over a cross-section. The third term should be zero because it includes the time average of the product of three quantities ρ', h', and u', oscillating with time. We call the resulting term *enthalpy flux density H* of acoustic wave in a tube, which is written as

$$H = \rho_m\langle\langle h'u'\rangle\rangle. \tag{4.16}$$

The enthalpy flux density is a vector quantity whose direction is represented by the sign; positive (negative) H means the energy flux density going in the positive (negative) direction of x.

The enthalpy fluctuation h' is expanded in terms of pressure fluctuation p' and entropy fluctuation S' as[2]

$$h' = \frac{1}{\rho_m}p' + T_m S'. \tag{4.17}$$

Therefore, the enthalpy flux density is decomposed as

$$H = I + Q, \tag{4.18}$$

---

[2]In a conventional description in thermodynamics, it is expressed as $dh = T dS + p dV$.

where I is the *work flux density* given by

$$I = \langle\langle p'u'\rangle\rangle \tag{4.19}$$

and Q is the *heat flux density* expressed by

$$Q = \rho_m T_m \langle\langle S'u'\rangle\rangle. \tag{4.20}$$

The work flux density I represents the acoustic power passing through a unit area, and is also called the acoustic intensity in acoustics, as we have introduced in Chapter 3. The heat flux density Q is also written as a product of the mean temperature T_m and the *entropy flux density* s;

$$Q = T_m s \tag{4.21}$$

where

$$s = \rho_m \langle\langle S'u'\rangle\rangle. \tag{4.22}$$

The entropy flux density s represents the hydrodynamic transport of entropy associated with the oscillatory motion of the gas. In adiabatic sound waves, $s = 0$ and hence $Q = 0$ because the entropy fluctuation of the gas satisfies $S' = 0$, although non-zero I may exist. In other words, the heat flux density is inherent to the acoustic waves in the pipe. When $S' \neq 0$ under the influence of thermal interaction between the gas and the tube wall, non-zero heat flux density Q can exist even if the steady mass flux density is zero. It was after the development of thermoacoustic devices that the importance of Q came to be recognized.

4.4.2 *Heat flow and work flow*

The total energy flow $\widetilde{H}$ in the axial direction is a sum of *work flow* $\widetilde{I}$ and *heat flow* $\widetilde{Q}$:

$$\widetilde{H} = \widetilde{I} + \widetilde{Q}. \tag{4.23}$$

If the gas-occupied cross-sectional area of the pipe is denoted by A, $\tilde{I}$ and $\tilde{Q}$ are respectively expressed as

$$\tilde{I} = AI, \tag{4.24}$$

$$\tilde{Q} = AQ. \tag{4.25}$$

In the particular case where conduction heat $\widetilde{Q}_\kappa$ by the wall and the gas should be considered, $\tilde{Q}$ is replaced with

$$\tilde{Q} = AQ + \tilde{Q}_\kappa. \tag{4.26}$$

When the pipe is thermally insulated, the axial energy flow is conserved, and hence

$$\frac{d\tilde{H}}{dx} = 0 \tag{4.27}$$

is satisfied. It is transformed as

$$\frac{d\tilde{I}}{dx} + \frac{d\tilde{Q}}{dx} = 0. \tag{4.28}$$

This equation means that the decrease of the work flow $\tilde{I}$ is compensated for by the increase of the heat flow $\tilde{Q}$. In other words, it represents the mutual energy conversion between $\tilde{I}$ and $\tilde{Q}$. The magnitude of the energy conversion per unit volume is called the *work source w*, which is given by

$$w = \frac{dI}{dx}. \tag{4.29}$$

Chapter 6 describes how the work source w depends on the oscillatory dynamics of the gas.

The ratio of heat flow $\tilde{Q}$ and temperature T_m means the entropy flow $\tilde{s}$:

$$\tilde{s} = \frac{\tilde{Q}}{T_m}. \tag{4.30}$$

The second law of thermodynamics states

$$\frac{d\tilde{s}}{dx} \geq 0, \tag{4.31}$$

which means that the entropy flow $\tilde{s}$ never decreases as it goes down. The *entropy production per unit volume*, σ_s, is given by

$$\sigma_s = \frac{ds}{dx} \geq 0. \tag{4.32}$$

4.4.3 Heat engine diagram using heat flow and work flow

The schematic diagram of the prime mover in Fig. 4.2 is redrawn in Figs. 4.5(a) to (c) by using the energy flows $\tilde{Q}$ and $\tilde{I}$. The energy flows go through the prime mover located between the hot and cold heat baths, changing their flow rates as a result of mutual energy conversions. Three types of prime movers are possible depending on their flow directions. In type (a), the work flow $\tilde{I}$ is amplified as it flows from cold to hot. This type of flow pattern is characteristic of a traveling-wave engine. In type (c), $\tilde{I}$ has different flow directions at ends. This type of prime mover corresponds to a standing-wave engine. In type (b), $\tilde{I}$ goes from hot to cold in the same way as $\tilde{Q}$. This type has not yet been found so far. In all the types, the heat flow $\tilde{Q}$ goes from hot to cold through the prime mover, decreasing its flow rate. The entropy flow $\tilde{s} = \tilde{Q}/T$ also goes through the prime mover.

If we integrate Eq. (4.28) from cold to hot through the prime mover, the following relation is obtained:

$$\Delta\tilde{I} + \Delta\tilde{Q} = 0,$$

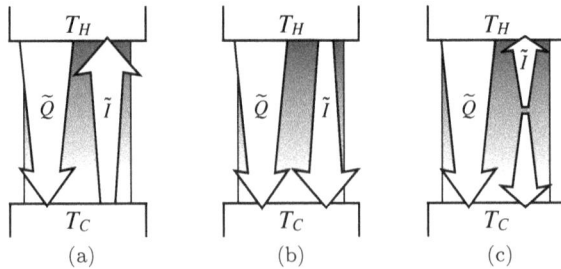

Figure 4.5: Schematic diagrams of prime mover.

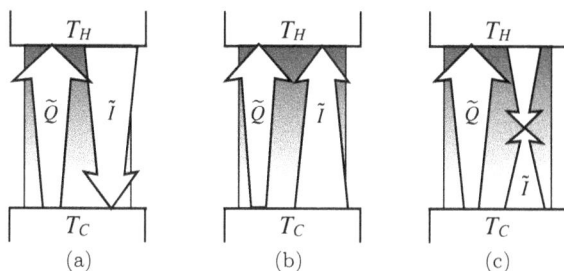

Figure 4.6: Schematic diagrams of heat pump.

where

$$\Delta \tilde{I} = \int_C^H \frac{d\tilde{I}}{dx} \, dx$$

$$\Delta \tilde{Q} = \int_C^H \frac{d\tilde{Q}}{dx} \, dx.$$

We should note that $\Delta \tilde{I}$ corresponds to work W in the first law of thermodynamics $W = Q_H - Q_C$.

Figure 4.6 illustrates schematic diagrams of heat pump. In heat pumps, heat flow goes from cold to hot. Depending on the flow pattern, heat pumps are classified into three types. Type (a) is typical of a looped tube cooler and pulse tube cooler (see Section 4.7), whereas type (b) can be realized in a resonance tube cooler [50]. The flow pattern in type (c) may be achieved in an acoustic cooler with a dual driver setting [58].

The heat engines in Figs. 4.5 and 4.6 are more easily and quantitatively illustrated by introducing an energy flow diagram in the next section.

4.5   Energy Flow Diagrams

4.5.1   *How to draw energy flow diagrams*

The energy flow diagram shows the flow rates of $\tilde{I}$ and $\tilde{Q}$ as a function of axial coordinate x. The flow direction is shown by its sign; a positive flow rate means the flow going to $+x$ direction, whereas

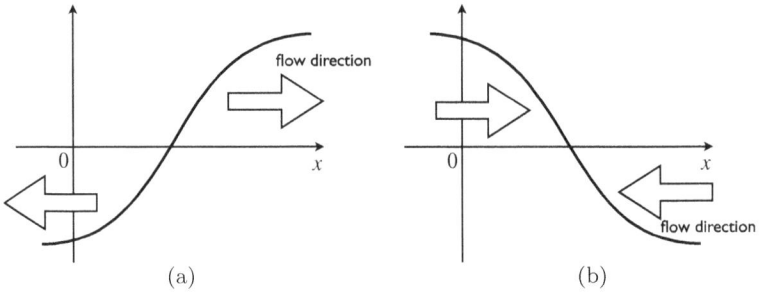

Figure 4.7: Energy flow diagram showing (a) source and (b) sink. The arrows indicate flow directions.

the negative one corresponds to $-x$ direction. When the flow rate has a positive (negative) slope, it means that the flow rate increases (decreases) as it flows down. Figure 4.7 shows a source and a sink where energy flow rates cross zero with positive and negative slopes, respectively. The energy flow is created in the source, while it is absorbed in the sink.

4.5.2 *Energy flow diagrams of prime movers*

Figure 4.8 represents the energy flow diagrams of the prime movers depicted in Fig. 4.5. Axial coordinate x is directed from cold to hot through the prime mover consisting of a regenerator, hot heat exchanger, cold heat exchanger, and tubes. The regenerator region ($b \leq x \leq c$) is assumed to be thermally insulated from the surrounding. Thus, the total energy flow $\tilde{I} + \tilde{Q}$ is kept the same, although $\tilde{I}$ and $\tilde{Q}$ may change their flow rates in the regenerator. The rate of energy conversion per unit time, W, is read off from the diagram as it is equal to the increase of $\tilde{I}$;

$$W = \int_b^c \frac{d\tilde{I}}{dx} \, dx = \tilde{I}|_c - \tilde{I}|_b.$$

Heat Q_H goes into the prime mover from the outside through the hot heat exchanger ($c \leq x < d$) per unit time. It is expressed by a

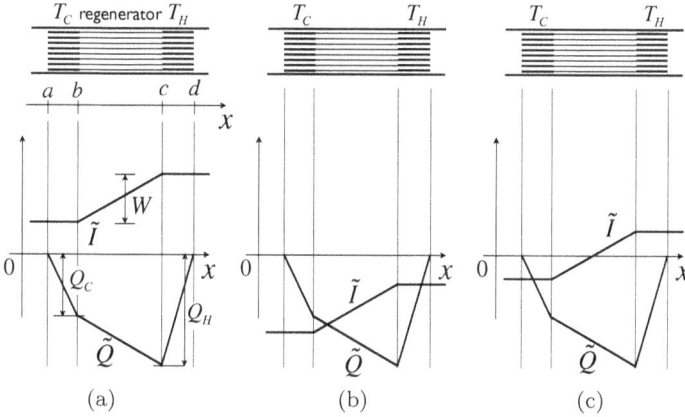

Figure 4.8: Energy flow diagrams of prime movers shown in Fig. 4.5.

change of $\widetilde{Q}$ in the hot heat exchanger;

$$Q_H = \int_c^d \frac{d\widetilde{Q}}{dx}\, dx = \widetilde{Q}|_d - \widetilde{Q}|_c.$$

If the tubes $(x < a,\ x > d)$ are so wide that $\omega\tau_\alpha \gg 1$ is satisfied, the heat flow $\widetilde{Q}$ is negligibly small. Therefore,

$$Q_H = -\widetilde{Q}|_c.$$

Heat Q_C goes out of the prime mover through the cold heat exchanger $(a \le x < b)$ region. It is equal to the reduction of $\widetilde{Q}$ in the cold heat exchanger, and hence,

$$Q_C = -\int_a^b \frac{d\widetilde{Q}}{dx}\, dx = \widetilde{Q}|_a - \widetilde{Q}|_b = -\widetilde{Q}|_b.$$

If we integrate Eq. (4.28) throughout the regenerator, we have

$$\int_b^c \frac{d\widetilde{I}}{dx}\, dx + \int_b^c \frac{d\widetilde{Q}}{dx}\, dx = \widetilde{I}|_c - \widetilde{I}|_b + \widetilde{Q}|_c - \widetilde{Q}|_b = 0.$$

This equation means the first law of thermodynamics: $W = Q_H - Q_C$.

In the same way as the energy flow diagram, one can draw an entropy flow diagram. Figure 4.9 represents the entropy flow diagram

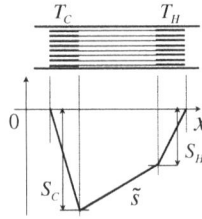

Figure 4.9: Entropy flow diagram of a prime mover.

of the prime mover. The entropy flow $\tilde{s}$ goes in the same direction as $\widetilde{Q}$. The slope $d\tilde{s}/dx$ means the entropy production per unit length, which never becomes negative. If we integrate Eq. (4.31) throughout the regenerator, we have

$$\int_b^c \frac{d\tilde{s}}{dx} \, dx = \tilde{s}|_c - \tilde{s}|_b \geq 0,$$

which is equivalent to the second law of thermodynamics: $S_C - S_H \geq 0$.

When Figs. 4.8 and 4.9 are compared with conventional diagram of the prime mover in Fig. 4.2, one can point out two differences: where and how the energy conversion takes place can be addressed by the work source $w = d\widetilde{I}/dx$, and where and how the entropy production takes place can be shown by the entropy production per unit length $d\tilde{s}/dx$. Spatial description by the energy flow and entropy flow diagrams would be useful in understanding the prime movers both locally and globally.

4.5.3 *Energy flow diagram for an ideal regenerator*

Figure 4.10 presents the energy flow diagrams and entropy flow diagrams for the regenerators of prime mover and heat pump when they achieve Carnot efficiency using an ideal gas as the working fluid. The axial coordinate x is directed in such a way that the entropy flow goes in the negative direction of x. Because of the absence of irreversible processes, the entropy flow satisfies $d\tilde{s}/dx = 0$, meaning that the flow rates are conserved. As heat flow $\widetilde{Q}$ is expressed by $\widetilde{Q} = T_m \tilde{s}$, the ratio of the flow rates at the hot end to that at

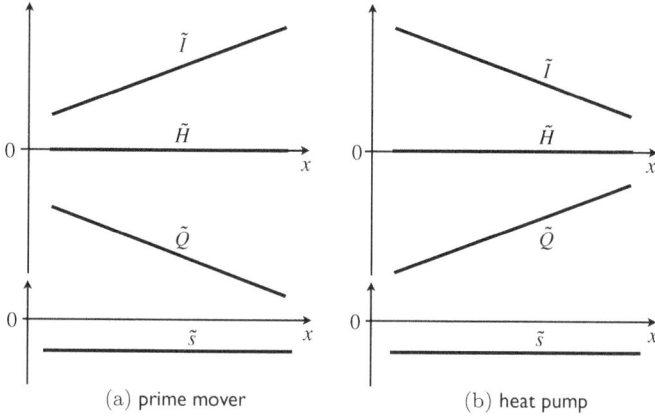

Figure 4.10: Energy flow diagram and entropy flow diagram of ideal prime mover and heat pump.

the cold end is equal to the temperature ratio T_H/T_C. The flow direction of $\widetilde{Q}$ is the same as $\tilde{s}$. In response to the flow rate change of $\widetilde{Q}$, the work flow $\widetilde{I}$ also changes the flow rate to satisfy the first law of thermodynamics: $d(\widetilde{I} + \widetilde{Q})/dx = 0$. We should note that the total energy flow rate satisfies $\widetilde{I} + \widetilde{Q} = 0$ in Fig. 4.10. The reason is explained as follows.

In the ideal regenerator where irreversible processes are not involved, the energy dissipation due to viscosity should be negligibly small, and more importantly, the heat exchange between the gas and the flow channel walls should take place isothermally. Therefore, the gas temperature at any location maintains its value: $T' = 0$. When the working gas is an ideal gas, the oscillating part of the enthalpy, h', is written as $h' = C_p T'$. The enthalpy flow $\widetilde{H}$ is then written as $\widetilde{H} = A\rho_m C_p \langle\langle T'u'\rangle\rangle = 0$. Because $\widetilde{H}$ is the total energy flow ($\widetilde{H} = \widetilde{I} + \widetilde{Q}$), $\widetilde{I} + \widetilde{Q} = 0$ should hold as in Fig. 4.10.

4.6 Examples of Energy Flow Diagrams

We introduce here some examples of energy flow diagrams in actual systems that are treated in this book. We should note that the measurement methods of $\widetilde{H}$ and $\widetilde{Q}$ have not yet been established, and therefore, the experimental results are available only for $\tilde{I}$. It is an urgent task to develop the measurement method of $\tilde{H}$ and $\tilde{Q}$.

4.6.1 Heat conduction

Figure 4.11 presents the energy flow diagram of a solid bar when its side wall is insulated from the environment. The bar is placed along x axis, and its end temperatures are maintained at T_H and T_C. The conducting heat $\tilde{Q}_\kappa$ is given by

$$\tilde{Q}_\kappa = -\kappa A_\kappa \frac{dT}{dx},$$

whereas the work flow is absent. Therefore, the energy conservation law states

$$\frac{d\tilde{Q}_\kappa}{dx} = 0,$$

and thus, the temperature T should change linearly with x if κ is independent of T.

The entropy production $d\tilde{s}/dx$ per unit length is given by

$$\frac{d\tilde{s}}{dx} = \frac{d}{dx}\left(\frac{\tilde{Q}_\kappa}{T}\right) = \frac{1}{T}\frac{d\tilde{Q}_\kappa}{dx} - \frac{\tilde{Q}_\kappa}{T^2}\frac{dT}{dx} = \kappa A_\kappa \left(\frac{1}{T}\frac{dT}{dx}\right)^2.$$

From the second law of thermodynamics, $d\tilde{s}/dx$ must be positive. Therefore, the thermal conductivity κ is positive in any material.

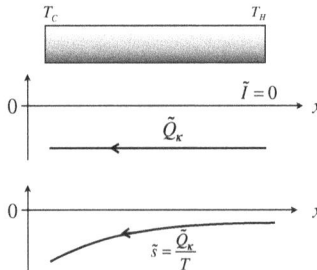

Figure 4.11: Energy flow diagram of a solid bar with temperature gradient.

4.6.2 *Adiabatic sound waves*

Consider here adiabatic sound waves of a gas column confined in a *wide tube* with $\omega \tau_\alpha \gg 1$. Because the gas particles adiabatically oscillate, we are able to assume $S = 0$. Therefore, only the work flow $\tilde{I}$ is present in the adiabatic sound waves. When the gas particles are free from energy dissipations at the internal surface of the tube wall, the energy conservation law holds:

$$\frac{d\tilde{I}}{dx} = 0. \tag{4.33}$$

Therefore, as shown in Fig. 4.12, $\tilde{I}$ maintains its flow rate. In thermoacoustic devices introduced in Chapter 1, the resonance tubes and the looped tubes are always made of relatively wide tubes, because their role is to deliver acoustic powers with less energy dissipations.

4.6.3 *Resonance tube with temperature gradient*

Figure 4.13 illustrates a resonance tube filled with air at ambient pressure. The tube contains a stack, which is sandwiched by two heat exchangers to provide a temperature gradient with it. The cold heat exchanger is kept at room temperature by a cooling water circuit, whereas the hot heat exchanger is heated by an electrical heater wire wound around it. When the temperature difference between the two heat exchangers goes beyond 191 K, the gas column starts to oscillate spontaneously with a frequency of 120.5 Hz, corresponding to the fundamental acoustic oscillation mode of the gas column.

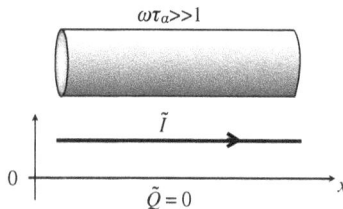

Figure 4.12: Energy flow diagram of adiabatic sound waves in a tube.

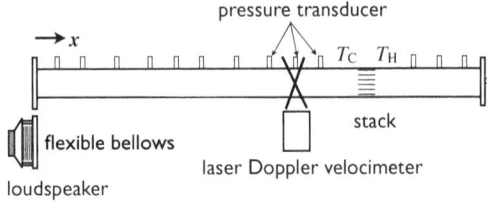

Figure 4.13: Experimental setup.

The generation of spontaneous gas oscillations can be analyzed from the energetic point of view through measurements of $\tilde{I}$ with various temperature ratio T_H/T_C of the two heat exchangers.

In order to measure $\tilde{I}$ with T_H/T_C less than the critical one, the end plate of the tube was replaced with an acoustic driver made of a woofer speaker and a dynamic bellows. By using the acoustic driver, the gas column was vibrated at the fundamental frequency, and also the acoustic pressure amplitude at the other end was maintained at 2.5×10^3 Pa, regardless of the temperature gradient along the stack. The work flux density $I = \langle\langle p'u'\rangle\rangle$ is transformed as

$$I = \frac{1}{2}\mathrm{Re}[p_1 u_{1r}^{\dagger}] \qquad (4.34)$$

by taking the cross-sectional average before the the time average, and also by taking into consideration that the pressure is independent of r. The cross-sectional mean velocity u_{1r} was determined from the velocity $u_1(0)$ on the central axis of the tube ($r = 0$) by

$$u_{1r} = \frac{1 - \chi_\nu}{1 - f_\nu(0)} u_1(0). \qquad (4.35)$$

The velocity u_1 was measured simultaneously with the pressure p_1 using the laser Doppler velocimeter and a pressure transducer.

Figure 4.14 summarizes the experimental results of I. When the stack temperature was uniform with $T_H/T_C = 1$, I was positive throughout the resonator. As positive I means the flow from left to right, we see that I is supplied from the acoustic driver at the left end and it flows down the resonance tube. The total amount of power supply can be evaluated by a product of the cross-sectional area of

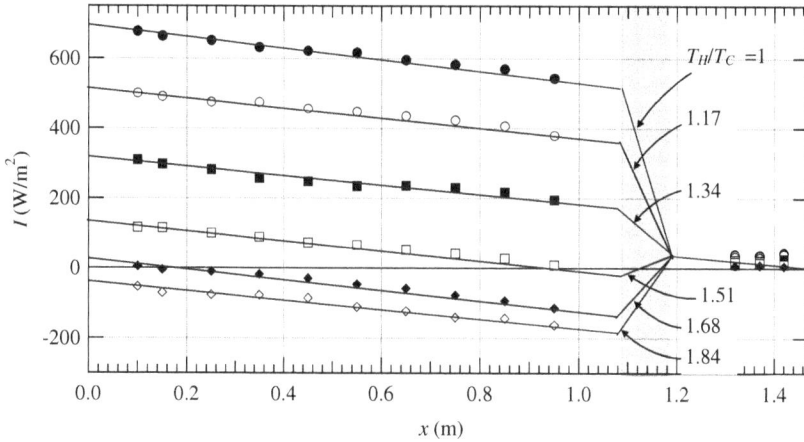

Figure 4.14: Axial distribution of work flux density I.

the tube and I at the left end. The decrease of I per unit length means the energy dissipation per unit volume. The larger slope in the stack region originates from viscous damping due to much narrower channel size than the resonance tube diameter.

As T_H/T_C becomes greater than unity, the slope of I in the stack region changes its sign from negative to positive. This result means that the energy conversion from Q to I becomes large enough to overcome the energy dissipation in the stack when T_H/T_C is sufficiently high. Correspondingly, I in the left side of the stack decreases, but the energy conversion is not large enough at $T_H/T_C = 1.68$ to cover the entire region of the tube. With $T_H/T_C = 1.84$, the energy conversion becomes so large that I flows out of the left end of the resonance tube. In other words, the resonance tube can supply more than enough acoustic power at this temperature ratio. Therefore, the acoustic driver is no longer necessary to maintain oscillations in the tube, because it serves as an acoustic load when $T_H/T_C = 1.84$. If the driver is replaced with a hard end, the acoustic amplitude would become more than 2.5×10^3 Pa because of the excess acoustic power. From this experimental result, we can say that the thermoacoustic oscillations occur because the stack serves as an acoustic power source when the temperature ratio is sufficiently large.

4.6.4 *Resonance tube engine and looped tube engine*

Through measurements of axial distribution of I, it has been clarified why the looped tube engine is essentially superior to the resonance tube engine. Yazaki *et al.* built a resonance tube engine and a looped tube engine by using the same components [27]. The only difference between them is whether the rigid plate is installed to block the acoustic wave transmission through it, as shown in Fig. 4.15(a). In the looped tube engine without the rigid plate, I goes around the loop in the direction from cold to hot through the stack. In the resonance tube engine with the rigid plate, on the other hand, I is emitted from both sides of the stack. As already shown in Fig. 4.5, the flow pattern of I in the stack region distinguishes the engine type; a traveling wave engine amplifies I in the stack, whereas a standing wave engine emits

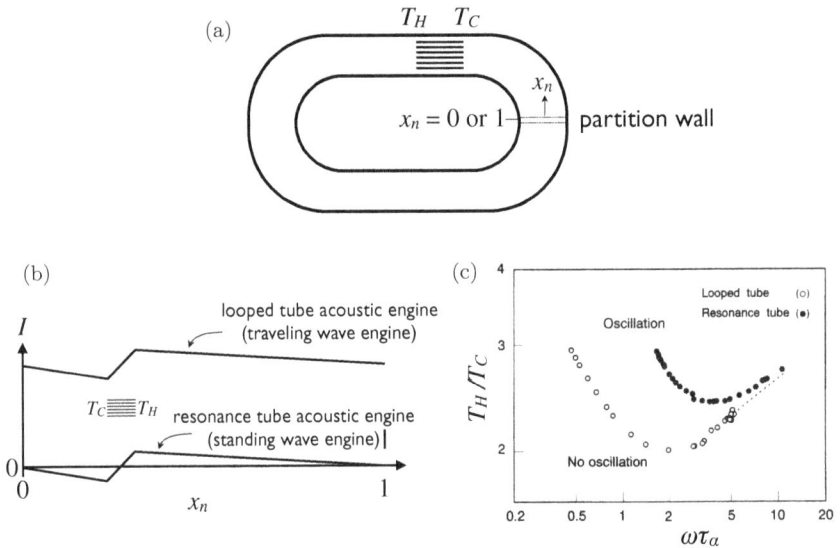

Figure 4.15: Resonance tube engine and looped tube engine. The looped tube engine in (a) changes to the resonance tube engine by inserting a rigid plate in the tube [27]. Axial distribution of I in (b) shows an essential difference of the stack between them; the stack plays a role of an amplifier of I in the looped tube engine, whereas it acts as a source of I in the resonance tube engine. The stability curve in (c) reflects the difference of the energy conversion mechanisms in the two types of engines.

I from both sides of the stack. These engine names originate from the phasing between pressure and velocity in the stack region, as will be explained in Chapter 6.

As shown in Fig. 4.10, an ideal engine amplifies I with a gain of T_H/T_C. Therefore, one can imagine that the looped tube engine has a better energy conversion efficiency than the resonance tube engine. The stability curve of the two engines shows that the looped tube engine starts to operate with a lower temperature ratio than the resonance tube engine. This result reflects the greater potential ability of the looped tube engine than the resonance tube engine.

4.6.5 *Stirling engine*

An α-type Stirling engine is comprised of a regenerator sandwiched by hot and cold heat exchanger and opposed pistons. The two pistons oscillate with the same frequency but with a finite phase difference about $90°$. Let us investigate the meaning of this phase difference from thermoacoustical point of view.

The model Stirling engine built for this experiment uses loud-speakers in place of solid pistons as shown in Fig. 4.16. The loudspeakers are placed at the sides of a 0.5-m long cylindrical pipe that contains air at atmospheric pressure and temperature as a working gas. They are driven by a two-channel function generator via

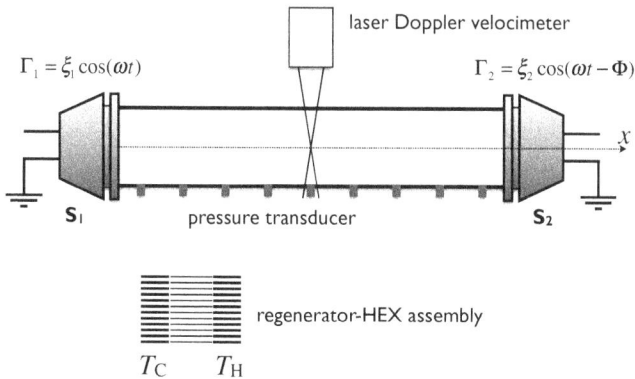

Figure 4.16: Model Stirling engine.

power amplifiers to oscillate at an angular frequency ω with a phase difference Φ between the driving voltages. Therefore, if we express the displacement oscillation of the left loudspeaker by $\Gamma_1 = \xi_1 \cos(\omega t)$, that of the right loudspeaker is written as $\Gamma_2 = \xi_2 \cos(\omega t - \Phi)$. The angular frequency ω is much lower than the natural frequencies of longitudinal acoustic oscillations of the gas in the pipe. Absence of the regenerator and two heat exchangers are intended to better observe the acoustic field created by the loudspeakers in the pipe. We show below the work flux density I when Φ was tuned to $0°$, $90°$, and $180°$ when keeping the same displacement amplitudes: $\xi_1 = \xi_2$.

Figures 4.17 and 4.18 respectively present I with $\Phi = 180°$, $0°$, and $90°$ as a function of axial coordinate x normalized with respect to the total length of the tube. With the phase differences $\Phi = 180°$ and $0°$, the work flux density I is markedly smaller than that with $\Phi = 90°$. In other words, the phasing of $\Phi = 90°$ is responsible for creating a larger I going in one direction at a given displacement amplitude. The slope of I is always negative, reflecting the energy

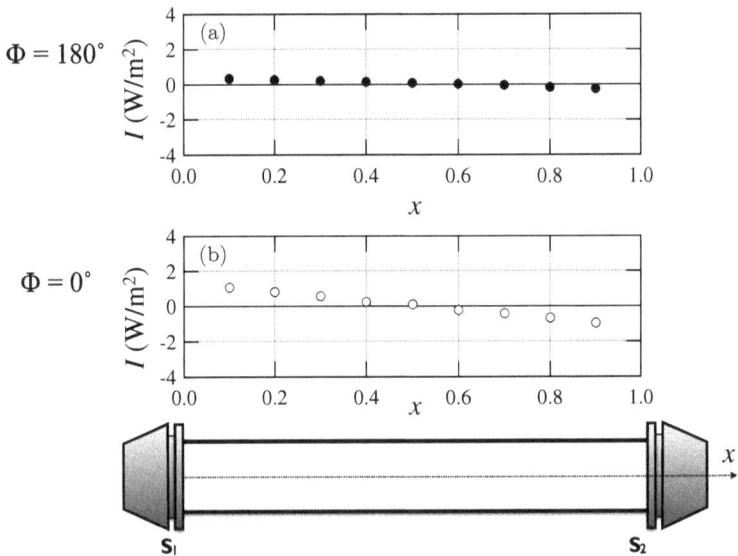

Figure 4.17: Work flux density in model Stirling engine: (a) $\Phi = 180°$ and (b) $\Phi = 0°$.

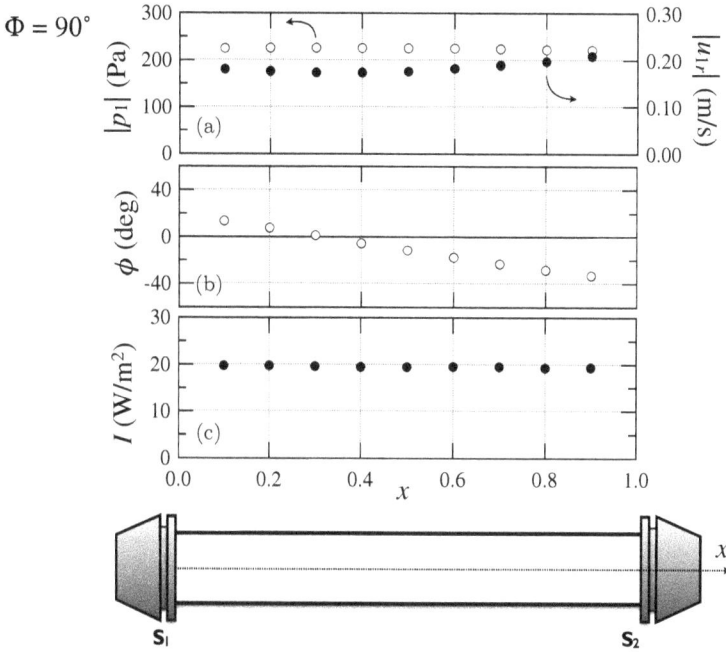

Figure 4.18: Acoustic field when $\Phi = 90°$: (a) pressure amplitude and amplitude of cross-sectional velocity, (b) phase difference between p and u_{1r}, (c) work flux density.

dissipation in the pipe. This phasing also results in a traveling wave phasing between pressure and velocity; as shown in Fig. 4.18(b), the phase lead ϕ of the velocity relative to the pressure is close to zero. Therefore, we should be able to amplify I if we install a differentially heated regenerator in the pipe, as pointed out by Ceperley.

Figure 4.19 shows the results with $\Phi = 90°$ when the regenerator was heated to possess a positive temperature gradient. The hot end temperature was 564 K, whereas the cold end temperature was 293 K. The regenerator was a ceramic catalyst support having lots of square pores with sides of 0.8 mm. The value of $\omega\tau_\alpha$ is 0.9 at the averaged temperature $(T_H + T_C)/2$. As we see from Fig. 4.19(c), I increased as it passed through the regenerator. This result demonstrates that the Stirling engine is classified into the type (a) of Fig. 4.5. The increase $\Delta I = 13.1\,\mathrm{W/m^2}$ means that the work source w is

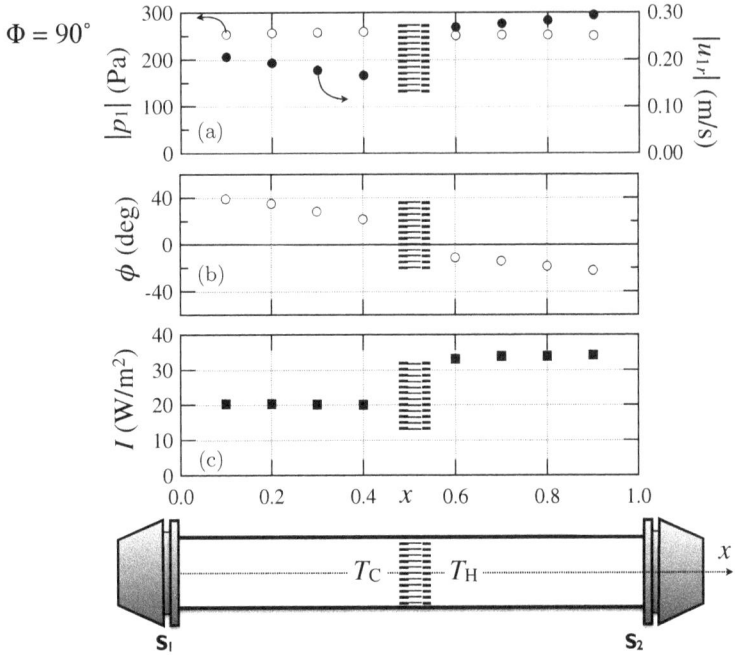

Figure 4.19: Acoustic field when $\Phi = 90°$: (a) pressure amplitude and amplitude of cross-sectional velocity, (b) phase difference between p and u_{1r}, (c) work flux density. Because of the temperature difference, I is amplified in the regenerator.

$0.7\,\mathrm{kW/m^3}$. Needless to say, this positive w is the result of the Stirling thermodynamic cycles as Ceperley predicted.

On the basis of the acoustic field obtained here, lets's make a comparison between the α-type Stiring engine and the looped tube acoustic engine in Fig. 4.20. In both engines, the heart of the engine is the regenerator, as it amplifies I through the Stirling cycles that the oscillating gas parcels execute. The part of I is consumed by the load and the rest is fed back to the regenerator. Only the difference between the two engines is seen in the feedback process of I. In the Stirling engine, two mechanically linked pistons convert the acoustic power received by the expansion piston to the shaft work and then convert it again to the acoustic power emitted from the compression piston; in the looped tube engine, on the other hand, the traveling acoustic wave carries I back to the cold end of the regenerator.

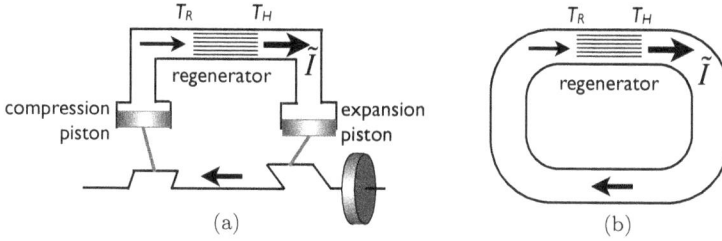

Figure 4.20: (a) α-type Stirling engine and (b) looped tube acoustic engine.

Therefore, the oscillating gas dynamics is essentially the same in the regenerators of the two engines.

4.7 Conceptual Design of Thermoacoustic Devices

The regenerator is central to thermoacoustic devices because the energy conversion between $\tilde{I}$ and $\tilde{Q}$ takes place there. In this section, let's try to understand the basic design of various thermoacoustic devices by considering $\tilde{I}$ and $\tilde{Q}$ that should promise better performance of the device.

4.7.1 *Pulse tube cooler and acoustic cooler*

A better cooling capacity can be achieved in an orifice pulse tube cooler compared to a basic pulse tube cooler. The reason is described here on the basis of energy flow. Figure 4.21(a) presents the schematic illustration and associated energy flows in a basic pulse tube cooler. The work flow $\tilde{I}_C$ at the cold end of the regenerator is small, because only a small amount of acoustic power is dissipated in the pulse tube made of a relatively wide tube. If an ideal regenerator is assumed for brevity, the relation $\tilde{Q} + \tilde{I} = 0$ should hold, therefore, the flow rate of $\tilde{Q}_C$ at the cold end also becomes small because $|\tilde{Q}_C| = |\tilde{I}_C|$.

The orifice pulse tube cooler has an orifice and a buffer tank next to the pulse tube. If the orifice can be regarded as a flow resistance, the energy dissipation readily becomes large owing to the gas flow entering the tank. As a result, $\tilde{I}_C$ and hence the cooling power $|\tilde{Q}_C|$ increase, compared to the basic pulse tube cooler. We should note

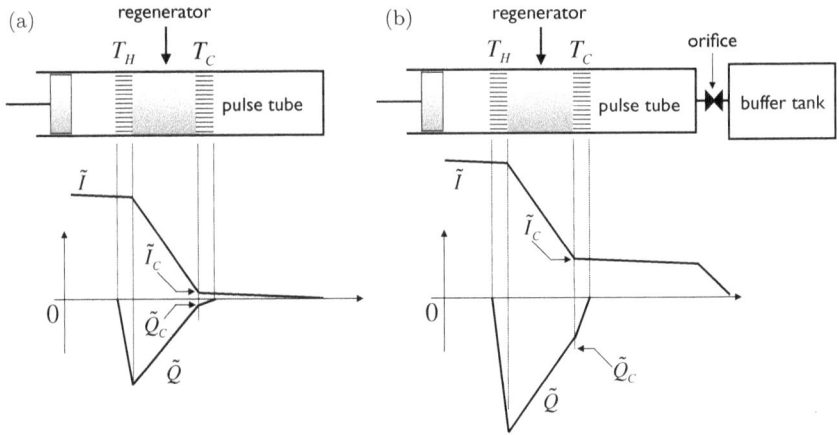

Figure 4.21: Schematic diagram of (a) Basic pulse tube cooler and (b) orifice pulse tube cooler, and associated energy flows.

that the larger cooling power in the orifice pulse tube cooler does not mean higher efficiency than the basic pulse tube cooler.

If we assume that the coefficient of performance, COP, is given by the ratio of cooling power over the work flow supplied to the cooler, it is expressed as $\text{COP} = |\widetilde{Q}_C/\widetilde{I}_H|$. In the ideal regenerator, we have seen that $|\widetilde{I}_C/\widetilde{I}_H| = T_C/T_H$ and also $\widetilde{Q}_C + \widetilde{I}_C = 0$ are satisfied in the previous section. Therefore, COP is given by

$$\text{COP} = \frac{T_C}{T_H}. \tag{4.36}$$

Although the ideal regenerator is assumed, COP is less than $\text{COP}_{\text{Carnot}} = T_C/(T_H - T_C)$. The lower COP is attributable to the fact that the cooling power is elevated by the dissipation at the orifice.

One way to solve this problem is to feed $\widetilde{I}$ going out of the regenerator back to the hot end by using a looped tube as shown in Fig. 4.22. If the energy dissipation in the looped tube is negligibly small, the work flow of the amount of $\widetilde{I}_H - \widetilde{I}_C$ is necessary for the acoustic driver at the end in the left-hand side of Fig. 4.22.

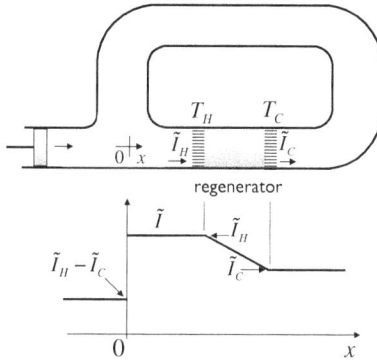

Figure 4.22: Schematic diagram and associated energy flows in a looped tube acoustic cooler.

Therefore, COP is expressed as

$$\mathrm{COP} = \left| \frac{\tilde{Q}_C}{\tilde{I}_H - \tilde{I}_C} \right|.$$

Because $\tilde{I}_H + \tilde{Q}_H = \tilde{I}_C + \tilde{Q}_C$ is satisfied in the ideal regenerator, the Carnot COP is achievable in the cooler with the feedback tube.

The ratio of COPs with and without the feedback tube is expressed by

$$\frac{T_C/T_H}{T_C/(T_H - T_C)} = 1 - \frac{T_C}{T_H}, \qquad (4.37)$$

which reaches unity with decreasing T_C/T_H. Thus, the absence of the feedback tube does not cause serious degradation of COP, for example in the cryocoolers intended to create liquid helium temperature. When T_C/T_H is close to 1, however, the ratio goes far below 1. Namely, a remarkable reduction of COP is unavoidable unless the feedback tube is attached to the cooler. If we consider the case with $T_C = 253\,\mathrm{K}$ $(-20°\mathrm{C})$ and $T_C = 300$ K $(27°\mathrm{C})$, T_C/T_H is as large as 0.84. Therefore, COP decreases down to 16% of $\mathrm{COP}_{\mathrm{Carnot}}$ in the orifice pulse tube refrigerator, even when the ideal regenerator is assumed.

4.7.2 Thermally driven acoustic cooler

A pistonless Stirling cooler, proposed in 2002 [39], has a prime-mover regenerator and a cooler regenerator in the same looped tube as shown in Fig. 4.23(a). This cooler is capable of making low temperatures without using any moving parts. This type of cooler is called a thermally-driven acoustic cooler. We consider here the thermal efficiency of the thermally-driven acoustic cooler by assuming ideal regenerators both for producing acoustic power and for pumping heat. The end temperatures of the prime-mover regenerator are T_H and T_R ($T_H > T_R$), and those for the cooler regenerator are T_R and T_C ($T_R > T_C$).

When the work flows at the cold and hot ends of the prime-mover regenerator are denoted by $\tilde{I}_R$ and $\tilde{I}_H$, respectively, we have $|\tilde{I}_H/\tilde{I}_R| = T_H/T_R$. Also, when those of the cooler regenerators are expressed by $\tilde{I}'_C$ and $\tilde{I}'_R$ respectively, we have a relation $|\tilde{I}'_C/\tilde{I}'_R| = T_C/T_R$. The thermal efficiency of the thermally-driven acoustic cooler

$$\epsilon = \left| \frac{\tilde{Q}'_C}{\tilde{Q}_H} \right| \tag{4.38}$$

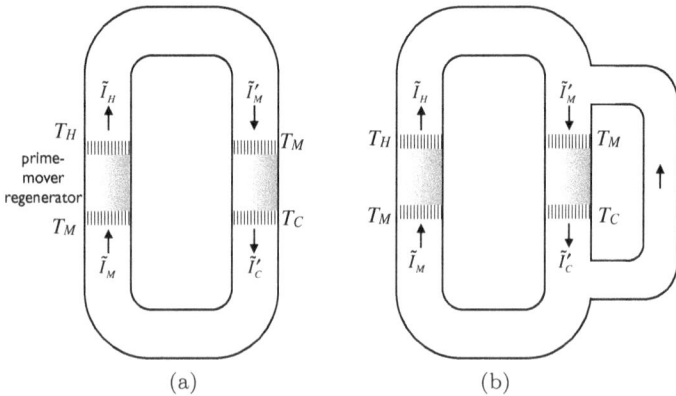

Figure 4.23: Thermally driven acoustic cooler: (a) single-loop configuration and (b) double-loop configuration.

is then transformed as follows:

$$\epsilon = \frac{\widetilde{I}'_C}{\widetilde{I}_H} = \frac{\dfrac{T_C}{T_R}\widetilde{I}'_R}{\dfrac{T_H}{T_R}\widetilde{I}_R} = \frac{T_C}{T_H}\frac{\widetilde{I}'_R}{\widetilde{I}_R}. \tag{4.39}$$

The relation between $\widetilde{I}_R$ and $\widetilde{I}'_R$ depends on how much energy dissipation is involved in the wide tube sections of the looped tube. If the energy dissipation is negligibly small, the work flow $\widetilde{I}_H$ going out of the prime-mover regenerator is equal to $\widetilde{I}'_R$ entering the cooler regenerator. Therefore, we have

$$\widetilde{I}'_R = \widetilde{I}_H = \frac{T_H}{T_R}\widetilde{I}_R. \tag{4.40}$$

In the same way, the work flow $\widetilde{I}'_C$ leaving the cooler regenerator is equal to $\widetilde{I}_R$ entering the prime-mover regenerator. Namely, we see

$$\widetilde{I}'_R = \frac{T_R}{T_C}\widetilde{I}_R. \tag{4.41}$$

These two equations are satisfied only when the temperature ratios are the same in the prime-mover regenerator and the cooler regenerator. The efficiency ϵ is then given by

$$\epsilon = \frac{T_R}{T_H} = \frac{T_C}{T_R}. \tag{4.42}$$

When the temperature ratios are not the same with each other, a continuous work flow is difficult to achieve. In other words, non-zero energy dissipation must be introduced when the single-loop configuration is adopted. In order to realize the thermally-driven cooler that can operate with arbitral temperature ratios, one should use a double-loop configuration as shown in Fig. 4.23(b) [59]. In this configuration, the thermal efficiency ϵ can be given for ideal regenerators as

$$\epsilon = \frac{T_H - T_R}{T_H}\frac{T_C}{T_R - T_C}, \tag{4.43}$$

which is a product of the Carnot efficiency for the prime mover and the Carnot COP for the cooler.

4.7.3　Acoustic engine having a series of regenerators

An ideal regenerator can function as an amplifier of the work flow with a gain of temperature ratio T_H/T_C of the regenerator ends:

$$G = \frac{T_H}{T_C}. \tag{4.44}$$

If a series of n regenerators are employed, the total gain becomes

$$G = \left(\frac{T_H}{T_C}\right)^n. \tag{4.45}$$

In the case of the three-stage amplifier shown in Fig. 4.24, the total gain of 8 is attainable with $T_H/T_C = 2$. When n is increased to 10, the total gain is raised to $G = 17$ with a relatively low temperature ratio of 1.33 when $T_H = 400\,\mathrm{K}$ and $T_C = 300$. If one tries to attain

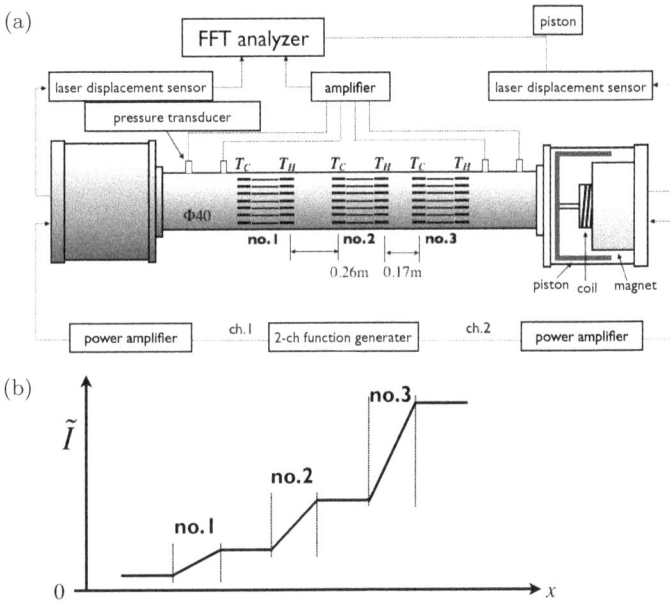

Figure 4.24: Acoustic power amplifier: (a) schematic illustration and (b) work flow.

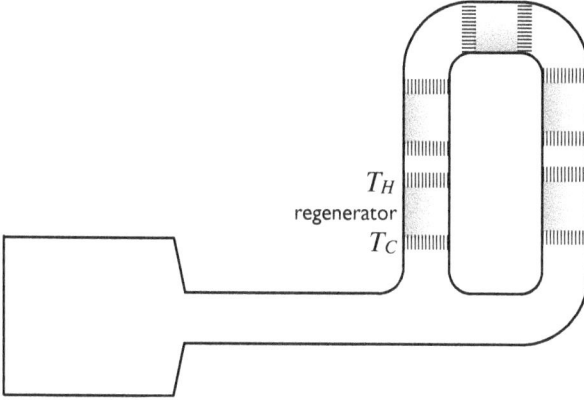

Figure 4.25: Low temperature differential acoustic engine using five regenerators.

$G = 17$ with a single regenerator, one needs T_H as high as 5100 K for $T_C = 300$ K. Therefore, the advantage of a series-configuration of the regenerator is clearly a capability of achieving a high gain [60, 61].

The larger gain leads to the larger output power $\Delta \widetilde{I}$ for a given temperature ratio T_H/T_C. The acoustic engine in Fig. 4.25 accommodates five regenerators in total. As a result, the critical temperature difference necessary for self-sustained oscillations is reduced to 51°C [62], whereas 225°C is required for the single regenerator case. Therefore, the acoustic engine with series regenerators would be useful for the applications in the field of exhaust heat recovery system, where the utilization of large scale heat sources with relatively low temperature are the target. Hasegawa *et al.* built such an engine using two looped tubes connected by a branch resonator [40]. Also de Blok has succeeded in reducing greatly the onset temperature difference by installing four regenerators in the looped tube acoustic engine [63].

We have illustrated the heat engines using energy flows and introduced the conceptual design of acoustic heat engines. In the next chapter, we explain the formulation of heat flow and work flow in a framework of hydrodynamics, and in Chapter 6, we present a more elaborate discussion on various acoustic engines and coolers.

4.8 Appendix

4.8.1 *Classification of natural phenomena based on energy flows*

Heat flow $\widetilde{Q}$ and entropy flow $\tilde{s}$ are related through $\widetilde{Q} = T_m\tilde{s}$, and therefore

$$\frac{d\widetilde{Q}}{dx} = \tilde{s}\frac{dT_m}{dx} + T_m\frac{d\tilde{s}}{dx}.$$

By inserting this relation into the first law of thermodynamics in Eq. (4.28), we have

$$\frac{d\widetilde{I}}{x} + \tilde{s}\frac{dT_m}{dx} = -T_m\frac{d\tilde{s}}{dx}.$$

From the second law of thermodynamics, we obtain

$$\frac{d\widetilde{I}}{dx} + \tilde{s}\frac{dT_m}{dx} \leq 0.$$

Consider a plane of dI/dx and $\tilde{s}(dT_m/dx)$ shown in Fig. 4.26 according to Tominaga's proposal. On this plane, a region with $d\widetilde{I}/dx + \tilde{s}(dT_m/dx) > 0$ contradicts the second law of thermodynamics. Any real phenomena should go into the region below it, and hence, the line $d\widetilde{I}/dx + \tilde{s}(dT_m/dx) = 0$ is a border line that divides a real world and another world.

An ideal engine without entropy production falls somewhere on that line. Usually dynamical systems are discussed in a uniform temperature with $dT_m/dx = 0$. Thus, such phenomena occupies the boundary between the second quadrant and the third quadrant. If the dissipation is ignored, the dynamical systems is bounded on the origin. Equilibrium thermodynamics treats a problem under dynamical equilibrium and thermal equilibrium, and therefore, $dT_m/dx = 0$ and $d\widetilde{I}/dx = 0$. So equilibrium thermodynamics is also bounded on the origin. A problem of thermal conduction is located somewhere on the vertical line. The prime mover is located in the fourth quadrant below the border line, whereas the cooler is placed in the second quadrant.

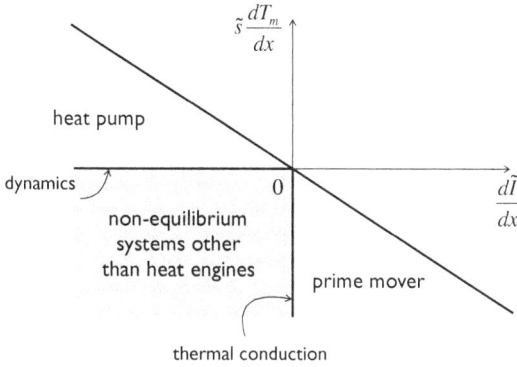

Figure 4.26: Classification of various phenomena. The border line is shown by a solid line with a negative slope, above which no real events would exist. Below the border line, heat pumps and prime movers occupy triangular regions on the left and right, whereas other events fall in the gray region.

The dream pipe goes into the second quadrant. The distance from the line $dT_m/dx = 0$ and $d\tilde{I}/dx = 0$ yields the measure of how far the heat engine is from the Carnot engine.

4.9 Problems

1 Show that the enthalpy flux density for an ideal gas is given by $H = \rho_m C_p \langle\langle T'u'\rangle\rangle$.

2 Draw schematically the diagrams of $\tilde{I}$ for the looped tube engine with a branch resonator in Fig. 4.27(a) and the heat driven acoustic cooler in Fig. 4.27(b).

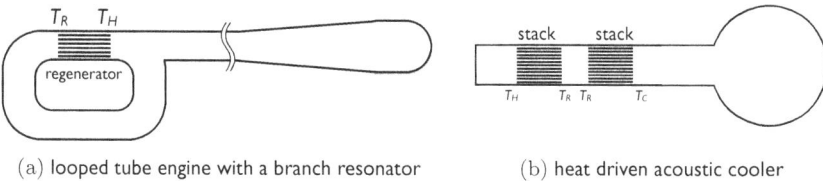

(a) looped tube engine with a branch resonator   (b) heat driven acoustic cooler

Figure 4.27: (a) looped tube engine with a branch resonator and (b) heat driven acoustic cooler.

Answers

1 The enthalpy flux is given by $H = \rho_m \langle\langle h'u' \rangle\rangle$. The fluctuating part of enthalpy, h', is expressed by $h' = T_m S' + \dfrac{1}{\rho_m}p'$. The fluctuating part of entropy, S' is also expressed by

$$S' = \left(\frac{\partial S}{\partial T}\right)_p T' + \left(\frac{\partial S}{\partial p}\right)_T p'$$

$$= \frac{C_p}{T_m}T' - \left(\frac{\partial V}{\partial T}\right)_p p'.$$

Because $\left(\dfrac{\partial V}{\partial T}\right)_p = \dfrac{1}{\rho_m T_m}$ is satisfied for the ideal gas, we obtain

$$h' = C_p T' + \left[\frac{1}{\rho_m} - T_m\left(\frac{\partial V}{\partial T}\right)_p\right]p' = C_p T'.$$

Therefore, H is given by $H = \rho_m \langle\langle h'u' \rangle\rangle$ for the ideal gas.

2 See the diagrams in Fig. 4.28.

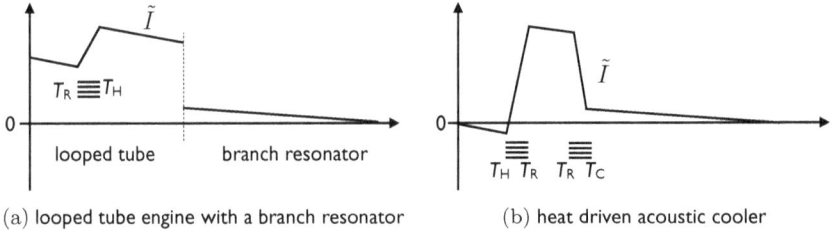

(a) looped tube engine with a branch resonator (b) heat driven acoustic cooler

Figure 4.28: Work flows in (a) looped tube engine with a branch resonator and (b) heat driven acoustic cooler.

Chapter 5

Basic Equations of Sound Waves in a Pipe and Their Solutions

Acoustic variables are described as Eulerian quantities in basic equations of hydrodynamics. By converting them to Lagrangian quantities, the energy conversion between heat flow and work flow is linked with thermodynamic cycles that are executed by each gas parcel. To investigate the thermodynamic cycles, the entropy oscillations of gas parcels are analyzed using the basic equations of hydrodynamics. The solution is also used to investigate the propagation constant of sound waves in a uniform temperature pipe. Hydrodynamics provides a basic understanding of thermoacoustic devices from a microscopic point of view.

5.1 Linearization of Hydrodynamic Equations

From conservation laws of mass, momentum, and energy, basic hydrodynamic equations are derived, as summarized at the end of this chapter. Equation of continuity is given as

$$\frac{\partial \rho}{\partial t} + \mathrm{div}(\rho \boldsymbol{u}) = 0. \tag{5.1}$$

Equation of motion (imcompressible Navier-Stokes equations) is given as

$$\rho \left(\frac{\partial \boldsymbol{u}}{\partial t} + (\boldsymbol{u} \cdot \nabla)\boldsymbol{u} \right) = -\nabla p + \mu \Delta \boldsymbol{u}, \tag{5.2}$$

where μ is the viscosity of the fluid. The general equation of heat transfer is expressed by

$$\rho T \left(\frac{\partial S}{\partial t} + (\boldsymbol{u} \cdot \nabla) S \right) = \kappa \Delta T + \Phi, \qquad (5.3)$$

where κ and Φ denotes the thermal conductivity and dissipation function, respectively. Throughout this book, μ and κ are constants independent of temperature, because relatively small amplitude oscillations are considered.

To describe thermoacoustic phenomena, we transform the hydrodynamic equations using two approximations. One is a linear approximation, where non-linear terms given by products of acoustic variables are ignored. This approximation allows simpler mathematical formulations, and moreover, makes the principle of superposition applicable. The other is a long wavelength approximation, where pipe diameter is sufficiently smaller than the wavelength. Owing to this approximation, the pressure waves are treated to have plane wavefront perpendicular to the tube axis.

In Chapter 2, sound waves propagating through a tube with uniform temperature were described. When temperature is constant along the tube axis, the mean density ρ_m and mean entropy S_m are constants. When temperature changes along the tube axis, however, ρ_m and S_m should be treated as a function of axial position x through x dependent mean temperature $T_m(x)$: $\rho_m = \rho_m(x)$ and $S_m = S_m(x)$. Through the gas viscosity and thermal conductivity, viscous and thermal interactions are introduced between the gas and the tube wall, which make the density fluctuation ρ' and entropy fluctuation S' dependent on a transversal coordinate r, as well as on x. Thus, the density and entropy of the gas are expressed as:

$$\rho(x, r, t) = \rho_m(x) + \rho'(x, r, t)$$
$$S(x, r, t) = S_m(x) + S'(x, r, t).$$

We assume that the temporal mean of velocity $\boldsymbol{u}$ is zero:

$$\boldsymbol{u}(x, r, t) = \boldsymbol{u}'(x, r, t).$$

The pressure is expressed as

$$p(x,t) = p_m + p'(x,t).$$

Here, r dependence is ignored because of the long-wavelength approximation, and p_m is constant so that the mean velocity is zero.

Equations (5.1)–(5.3) are generally difficult to solve, but they can be greatly simplified by linear approximation. For example, a second order term like $\rho' \boldsymbol{u}'$ is ignored with respect to $\rho_m \boldsymbol{u}'$.

The continuity equation in Eq. (5.1) is written as

$$\frac{\partial(\rho_m + \rho')}{\partial t} + \operatorname{div}[(\rho_m + \rho')\boldsymbol{u}] = 0.$$

If it is left with first order variables, the linearized continuity equation is obtained:

$$\frac{\partial \rho'}{\partial t} + u' \frac{d\rho_m}{dx} + \rho_m \frac{\partial u'}{\partial x} = 0 \tag{5.4}$$

where u' is the axial component of $\boldsymbol{u}'$.

The axial component of equation of motion is also rewritten as

$$\rho_m \frac{\partial u'}{\partial t} = -\frac{\partial p'}{\partial x} + \mu \Delta_\perp u'. \tag{5.5}$$

The operator $\Delta_\perp$ means the Laplace operator with respect to the cross-section of the tube, and is given in a cylindrical coordinate system as

$$\Delta_\perp = \frac{\partial^2}{\partial r^2} + \frac{1}{r}\frac{\partial}{\partial r}. \tag{5.6}$$

The linearized general equation of heat transfer is written in the same way as

$$\rho_m T_m \left(\frac{\partial S'}{\partial t} + \frac{dS_m}{dx}u'\right) = \kappa \Delta_\perp T'. \tag{5.7}$$

The linear approximation is validated only when the variables are sufficiently small, but its applicability is practically wide. Past studies have shown that the *high amplitude acoustic waves* refer to the case when pressure amplitude exceeds 10% of the mean pressure. In an adiabatic traveling wave, the corresponding velocity amplitude is about 10% of the speed of sound $[u/c_S = p/(\gamma p_m)]$. With this amplitude level, one cannot ignore various non-linear phenomena like shock wave formation. Consider the acoustic pressure waves of 10 kPa in amplitude propagating in air at ambient pressure (100 kPa). For an adiabatic traveling wave, the acoustic intensity is 122 kW/m^2, which corresponds to 960 W in a pipe of 100 mm in diameter. If the mean pressure is raised, the power range of the "small pressure oscillations" also increases. One can discuss a wide variety of thermoacoustic devices under the linear approximation.

5.2   Eulerian Description and Lagrangian Description

There are two approaches for describing the fluid motion: Eulerian approach and Lagrangian approach. In the former, the observer sees the fluid from a coordinate system fixed in space, and in the latter the observer watches the fluid while following its motion. The Eulerian description is mathematically simpler than the Lagrangian description. Therefore, the classical field theories like electromagnetic theory and fluid dynamics usually adopt the Eulerian description. Thermodynamics, on the other hand, focuses on a specific volume of fluid and describes what happens to it. So the Lagrangian description is more suitable if we put emphasis on discussion of energy conversions in thermoacoustic devices. In the following, we summarize the relation between the two approaches and derive the conversion formula.

Consider here an acoustic variable X that changes with position x and time t. The change of X in the Eulerian approach is described by the observer fixed in space, for example at $x = a$, as shown in Fig. 5.1. So, the change $\Delta_E X$ between time t and $t + \Delta t$ is expressed as

$$\Delta_E X = X(a, t + \Delta t) - X(a, t).$$

Also, consider a fluid parcel that moves along x with time. Suppose that the position of the gas parcel at time t is $x = a$. The observer moving with the gas parcel describes X a function of t. So, X would be expressed as $X(a; t)$ in the Lagrangian approach. Here a is used as a parameter to explicitly show which gas parcel is under consideration. If the gas parcel moves from a to $a + \Delta\xi$ after time Δt, the change $\Delta_L X = X(a; t + \Delta t) - X(a; t)$ in the Lagrangian approach is written as

$$\Delta_L X = X(a + \Delta\xi, t + \Delta t) - X(a, t).$$

Therefore, the relation between $\Delta_E X$ and $\Delta_L X$ is expressed as

$$\Delta_L X = \Delta_E X + X(a + \Delta\xi, t + \Delta t) - X(a, t + \Delta t).$$

For a small $\Delta\xi$, the term $X(a + \Delta\xi, t + \Delta t)$ on the right hand side can be expanded as

$$X(a + \Delta\xi, t + \Delta t) = X(a, t + \Delta t) + \Delta\xi \frac{\partial X(a, t + \Delta t)}{\partial x}$$
$$+ \frac{1}{2}(\Delta\xi)^2 \frac{\partial^2 X(a, t + \Delta t)}{\partial x^2} + \cdots$$

By ignoring the second and higher order terms, we arrive at

$$\Delta_L X = \Delta_E X + \Delta\xi \frac{\partial X(a, t + \Delta t)}{\partial x}.$$

As can be seen from Fig. 5.1, the first term on the right hand side expresses the change caused by the temporal change of the field $X(x, t)$, while the second term represents the change due to the

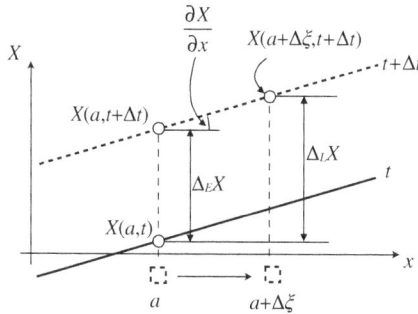

Figure 5.1: Axial distribution of a variable X at time t (solid line) and at time $t + \Delta t$ (dashed line). The change of X described in the Eulerian method is shown as $\Delta_E X$, while that in the Lagrangian method is shown as $\Delta_L X$. The dashed square represents the gas parcel under consideration, which moves from $x = a$ to $x = a + \Delta\xi$ after time Δt.

displacement in the spatially-dependent field. By dividing by time Δt, we obtain

$$\frac{\Delta_L X}{\Delta t} = \frac{\Delta_E X}{\Delta t} + \frac{\Delta \xi}{\Delta t} \frac{\partial X(a, t + \Delta t)}{\partial x}.$$

The time derivative of X in the Lagrangian method is therefore given by taking a limit of an infinitely small Δt as

$$\frac{dX(a;t)}{dt} = \frac{\partial X}{\partial t} + u \frac{\partial X(a,t)}{\partial x},$$

where the differential operator

$$\frac{d}{dt} = \frac{\partial}{\partial t} + u \frac{\partial}{\partial x} \tag{5.8}$$

is called Lagrangian derivative or material derivative. When X is expressed as a sum of the temporal mean X_m and the fluctuation X' ($X = X_m + X'$), the Lagrangian derivative of X' (see Fig. 5.10) is given by

$$\frac{dX'}{dt} = \frac{\partial X'}{\partial t} + \frac{dX_m}{dx} u'. \tag{5.9}$$

For a fluid particle whose mean position is $x = a$, $\partial X/\partial t$ and dX_m/dx in Eq. (5.9) are those at $x = a$.

By integrating Eq. (5.9), a change X' of X that a given fluid particle undergoes (Lagrangian method) is written as

$$X' + \frac{dX_m}{dx} \xi'$$

or in a complex form as

$$X_1 + \frac{dX_m}{dx} \xi_1$$

where the first term is the change of X observed at a fixed position (Eulerian method), and the second term represents the change of X caused by the displacement of the fluid parcel, ξ' denotes the displacement of the given fluid particle; X_1 and ξ_1 are the complex amplitudes. When a non-zero dX_m/dx exists, the change in the Lagrangian method is different from that of the Eulerian method. When a temperature gradient dT_m/dx is present, entropy $S_m(x)$ and density $\rho_m(x)$ depend on x, as well as $T_m(x)$. Therefore, we should notice that the associated changes of T', S', and ρ' are different in the Lagrangian method and the Eulerian method.

5.3 Lagrangian Representation of Energy Flux Density and Work Source

Work flux density I, heat flux density Q, and work source w have been introduced from the energy equation expressed in the Eulerian method in the following way.

$$I = \langle\langle p'u'\rangle\rangle, Q = \rho_m T_m \langle\langle S'u'\rangle\rangle, w = \frac{dI}{dx}.$$

Let us transform them in Lagrangian method.

5.3.1 *Work flux density*

Pressure fluctuation p' included in I is the same both in the Eulerian method and the Lagrangian method, as the mean pressure p_m satisfies $dp_m/dx = 0$. Therefore, the representation of I also becomes the same in both methods:

$$I = \langle\langle p'u'\rangle\rangle.$$

In the Lagrangian method, p' and u' are the pressure and the velocity of the gas particle under consideration.

5.3.2 *Heat flux density*

Heat flux density Q is expressed using entropy fluctuation S'. When the mean entropy S_m is not uniform over a space, S' in the Lagrangian method becomes different from that in the Eulerian method because of non-zero dS_m/dx. By expressing the Eulerian entropy fluctuation with the Lagrangian entropy fluctuation, we have

$$Q = \rho_m T_m \left\langle\left\langle \left(S' - \frac{dS_m}{dx}\xi'\right) u'\right\rangle\right\rangle$$

$$= \rho_m T_m \langle\langle S'u'\rangle\rangle - \rho_m T_m \frac{dS_m}{dx} \langle\langle \xi'u'\rangle\rangle.$$

We should note that $\langle \xi' u' \rangle_t = 0$ because of the relation $u' = d\xi'/dt$; thus, $\langle\langle \xi' u' \rangle\rangle = 0$ in the last expression. Therefore, heat flux density in the Lagrangian method is represented by

$$Q = \rho_m T_m \langle\langle S' u' \rangle\rangle$$

which is the same as in the Eulerian method. In the Lagrangian method, S' denotes the entropy fluctuation of the gas particle under consideration, and ρ_m and T_m are the temporal mean density and temperature of that gas particle.[1]

5.3.3 *Work source*

Work source w is expressed in the same way in the Eulerian method and the Lagrangian method, because p' and u' in I are common to both methods. Here we examine the physical meaning of w from a Lagrangian point of view.

By following the definition of $w = dI/dx$, we have

$$w = \left\langle\left\langle \frac{\partial p'}{\partial x} u' \right\rangle\right\rangle + \left\langle\left\langle p' \frac{\partial u'}{\partial x} \right\rangle\right\rangle. \qquad (5.10)$$

Inserting the equation of motion [Eq. (5.2)] into the first term on the right hand side yields

$$\left\langle\left\langle \frac{\partial p'}{\partial x} u' \right\rangle\right\rangle = \left\langle\left\langle \left(-\rho_m \frac{\partial u'}{\partial t} + \mu \Delta_\perp u' \right) u' \right\rangle\right\rangle = \langle\langle (\mu \Delta_\perp u')u' \rangle\rangle.$$

We have used the relation $\langle\langle (\partial u'/\partial t)u' \rangle\rangle = 0$ in the transformation above. In the Lagrangian point of view, $\mu \Delta_\perp u'$ can be interpreted as a viscous force that exerts on the gas particle of unit volume. Therefore, $(\mu \Delta_\perp u')u'$ represents the instantaneous viscous dissipation per unit volume and time. Namely,

$$W_\nu = \langle\langle (\mu \Delta_\perp u')u' \rangle\rangle$$

is the viscous dissipation per unit volume and time, as we have introduced in Chapter 3.

---

[1] $\rho_m = \rho_m(a)$ and $T_m = T_m(a)$ if the gas particle oscillates around $x = a$.

The second term on the right hand side of Eq. (5.10) is rewritten using the equation of continuity:

$$\frac{\partial u'}{\partial x} = -\frac{1}{\rho_m}\frac{d\rho'}{dt},\qquad(5.11)$$

where Eq. (5.1) is expressed using the time derivative of Lagrangian density fluctuation. By inserting this equation into the second term of Eq. (5.10) gives

$$\left\langle\!\left\langle p'\frac{\partial u'}{\partial x}\right\rangle\!\right\rangle = -\frac{1}{\rho_m}\left\langle\!\left\langle p'\frac{d\rho'}{dt}\right\rangle\!\right\rangle.$$

The right hand side of the equation above is transformed through the relation $\rho = 1/V$ between the density $\rho = \rho_m + \rho'$ and the volume $V = V_m + V'$:

$$\left\langle\!\left\langle p'\frac{\partial u'}{\partial x}\right\rangle\!\right\rangle = \rho_m\left\langle\!\left\langle p'\frac{dV'}{dt}\right\rangle\!\right\rangle.$$

Consequently, the work source w is given as

$$w = \rho_m\left\langle\!\left\langle p'\frac{dV'}{dt}\right\rangle\!\right\rangle + W_\nu.\qquad(5.12)$$

In Chapter 4, the output power resulting from the thermodynamic cycles was represented by

$$W = \oint p dV,$$

where p and V are associated with the whole working gas, for example, contained in a cylinder. On the other hand, Eq. (5.12) represents that the gas parcels individually function as small-scale heat engines. The thermodynamic cycles may be different from each other because the gas parcels occupy different positions in tiny flow channels of the regenerator shown in Fig. 5.2. Therefore, there is no unique thermodynamic cycle that characterizes a given thermoacoustic device; w is a spatially dependent quantity.

It is worth investigating the volume fluctuation V' of the given gas parcel in more detail. By expanding it in terms of pressure and

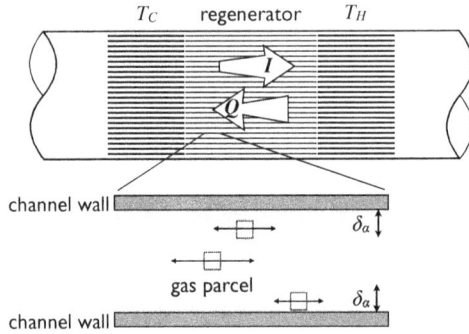

Figure 5.2: Microscopic view of gas parcels in a regenerator. The gas parcel oscillating near the wall surface experiences isothermal thermodynamic cycles, whereas those away from the thermal boundary layer tend to undergo adiabatic thermodynamic cycles.

entropy, we obtain

$$\frac{dV'}{dt} = \left(\frac{\partial V}{\partial p}\right)_S \frac{dp'}{dt} + \left(\frac{\partial V}{\partial S}\right)_p \frac{dS'}{dt}.$$

Since

$$\left\langle p' \frac{dp'}{dt} \right\rangle_t = 0$$

is always satisfied, the first term on the right hand side of Eq. (5.12) is rewritten as

$$\rho_m \left\langle \left\langle p' \frac{dV'}{dt} \right\rangle \right\rangle = \rho_m \left(\frac{\partial V}{\partial S}\right)_p \left\langle \left\langle p' \frac{dS'}{dt} \right\rangle \right\rangle.$$

Therefore, it is not the volume fluctuation $(\partial V/\partial p)_S p'$ due to adiabatic pressure change but $(\partial V/\partial S)_p S'$ due to isobaric entropy change that contributes to the work source. This result means that the gas parcels act as heat engines only when the thermally induced volume change is incorporated with the pressure change. Let us write the final form of $w = dI/dx$:

$$w = \rho_m \left(\frac{\partial V}{\partial S}\right)_p \left\langle \left\langle p' \frac{dS'}{dt} \right\rangle \right\rangle + W_\nu. \tag{5.13}$$

5.4 Equation Describing Entropy Fluctuation

The Lagrangian entropy fluctuation S' plays a key role both in the heat flux density Q and the work source w. In this section, how S' is related to p' and u' is investigated on the basis of a general equation of heat transfer in Eq. (5.7):

$$\rho_m T_m \left(\frac{\partial S'}{\partial t} + \frac{dS_m}{dx} u' \right) = \kappa \Delta_\perp T'.$$

Temperature fluctuation T' is expanded as

$$T' = \left(\frac{\partial T}{\partial p} \right)_S p' + \left(\frac{\partial T}{\partial S} \right)_p S'.$$

By taking into account the relation $\Delta_\perp p' = 0$ for a plane pressure wave, we have

$$\rho_m T_m \left(\frac{\partial S'}{\partial t} + \frac{dS_m}{dx} u' \right) = \kappa \left(\frac{\partial T}{\partial S} \right)_p \Delta_\perp S'.$$

The thermodynamic relation

$$\left(\frac{\partial T}{\partial S} \right)_p = \frac{T_m}{C_p}$$

allows us to further transform Eq. (5.7) as

$$\frac{\partial S'}{\partial t} + \frac{dS_m}{dx} u' = \alpha \Delta_\perp S', \tag{5.14}$$

where $\alpha = \kappa/(\rho_m C_p)$ denotes the thermal diffusivity of the gas. By using the complex amplitude of S' as S_1, we have

$$i\omega S_1 + \frac{dS_m}{dx} u_1 = \alpha \Delta_\perp S_1,$$

which can be transformed as

$$S_1 = \frac{\alpha}{i\omega} \Delta_\perp S_1 - \frac{dS_m}{dx} \frac{u_1}{i\omega}. \tag{5.15}$$

As shown in Eq. (3.83), the velocity amplitude u_1 is written by using the cross-sectional velocity amplitude $u_{1r} = \langle u_1 \rangle_r$ as

$$u_1 = \frac{1 - f_\nu}{1 - \chi_\nu} u_{1r}.$$

The entropy gradient dS_m/dx is expressed as

$$\frac{dS_m}{dx} = \left(\frac{\partial S}{\partial p}\right)_T \frac{dp_m}{dx} + \left(\frac{\partial S}{\partial T}\right)_p \frac{dT_m}{dx},$$

which is further transformed as

$$\frac{dS_m}{dx} = \left(\frac{\partial S}{\partial T}\right)_p \frac{dT_m}{dx}$$

because $dp_m/dx = 0$.

Thus, we obtain the following equation for S_1:

$$S_1 = \frac{\alpha}{i\omega}\Delta_\perp S_1 - \left(\frac{\partial S}{\partial T}\right)_p \frac{dT_m}{dx}\frac{1 - f_\nu}{1 - \chi_\nu}\frac{u_{1r}}{i\omega}. \tag{5.16}$$

The boundary condition for S_1 is given by

$$S_1|_{r=r_0} = \left(\frac{\partial S}{\partial p}\right)_T p_1|_{r=r_0} + \left(\frac{\partial S}{\partial T}\right)_p T_1|_{r=r_0}.$$

The pressure p_1 is uniform over the cross-section of the flow channel by the long wavelength approximation, and also suppose that the temperature at the wall surface is maintained at the wall temperature ($T_1|_{r=r_0} = 0$) because of relatively high heat capacity of the solid wall compared to the gas. The boundary condition is then simplified as

$$S_1|_{r=r_0} = S_1(r_0) = \left(\frac{\partial S}{\partial p}\right)_T p_1, \tag{5.17}$$

where $(\partial S/\partial p)_T p_1$ means the isothermal entropy fluctuation caused by pressure fluctuation.

5.5   Entropy Oscillations in the Absence of Axial Temperature Gradient

In the absence of axial temperature gradient $(dT_m/dx = 0)$, Eq. (5.16) is reduced to

$$S_1 = \frac{\alpha}{i\omega}\Delta_\perp S_1. \tag{5.18}$$

We assume a solution of the form

$$S_1 = f_\alpha(r)S_1(r_0) \tag{5.19}$$

where f_α is an unknown function of r. Inserting into the equation above yields

$$f_\alpha = \frac{\alpha}{i\omega}\Delta_\perp f_\alpha.$$

This differential equation is equivalent to that for f_ν introduced in Chapter 3. Therefore, the general solution is given by a sum of zeroth order Bessel function and zeroth order Neumann function, in the same way as f_ν. The boundary conditions for f_ν are given from Eq. (5.17) and also from the condition for the entropy fluctuation to be smoothly joined at $r = 0$:

$$f_\alpha|_{r=r_0} = 1, \quad \left.\frac{\partial f_\alpha}{\partial r}\right|_{r=0} = 0.$$

Namely, the boundary conditions for f_α are also the same as those for f_ν. Therefore, the solution is simply given by replacing ν with α as

$$f_\alpha = \frac{J_0(\eta_\alpha)}{J_0(\eta_{\alpha 0})} \tag{5.20}$$

$$\eta_\alpha = (i-1)\frac{r}{\delta_\alpha}, \quad \eta_{\alpha 0} = (i-1)\frac{r_0}{\delta_\alpha}, \quad \delta_\alpha = \sqrt{\frac{2\alpha}{\omega}}. \tag{5.21}$$

A dimensionless parameter $\omega\tau_\alpha$ is given by

$$\omega\tau_\alpha = \omega\frac{r_0^2}{2\alpha}. \tag{5.22}$$

Therefore, it is expressed by using the thermal boundary layer thickness δ_α as

$$\omega\tau_\alpha = \left(\frac{r_0}{\delta_\alpha}\right)^2.$$

Although the parameter $\omega\tau_\alpha$ was introduced rather intuitively in Chapter 2, it is shown here that the entropy oscillation is parametrized by $\omega\tau_\alpha$. It should be noted that the Eulerian entropy and the Lagrangian entropy have the same expression because $dT_m/dx = 0$.

5.5.1 *Radial profile of entropy fluctuation*

The function $f_\alpha(r)$ of the complex amplitude $S_1(r) = f_\alpha(r)S_1(r_0)$ is plotted in Fig. 5.3 as a function of the radial coordinate r for the cases with $\omega\tau_\alpha = 100, 10, 1,$ and 0.1. Figure 5.3(a) shows the

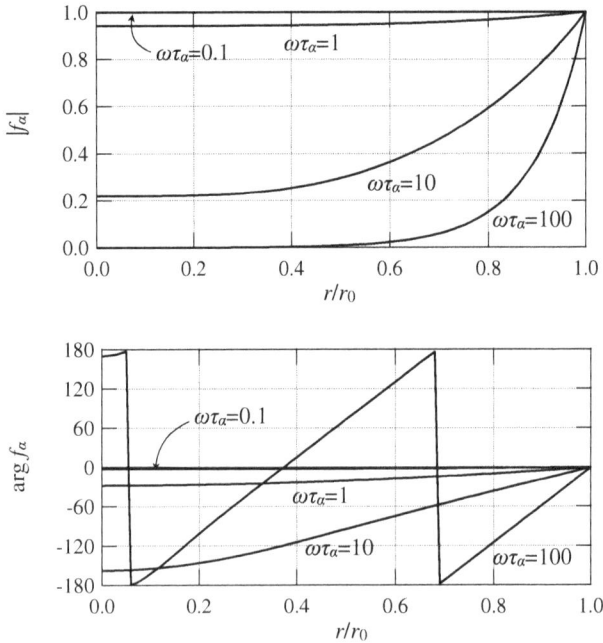

Figure 5.3: (a) Amplitude ratio $|f_\alpha| = S_1(r)/S_1(0)$ and (b) phase $\arg f_\alpha$ of $S_1(r)$ relative to $S_1(r_0)$ vs. r/r_0.

amplitude ratio $|f_\alpha|$ of $S_1(r)$ over $S_1(r_0)$, and (b) presents the phase $\arg f_\alpha$ relative to $S_1(r_0)$, where $S_1(r_0)$ denotes the isothermal entropy fluctuation at the internal wall surface $(r = r_0)$.

When $\omega\tau_\alpha$ is rather large $(\omega\tau_\alpha = 100)$, $|f_\alpha|$ rapidly decreases from unity at $r = r_0$, and becomes almost zero at $3\delta_\alpha$ away from the wall surface. We are able to say that the gas adiabatically oscillates in the core of the tube. When $\omega\tau_\alpha = 10$, the influence of the thermal boundary layer extends to the center of the tube. When $\omega\tau_\alpha = 1$, $|f_\alpha| = 0.94$ on the center, and finally with $\omega\tau_\alpha = 0.1$ $|f_\alpha| = 1$ is achieved almost entirely over the cross-section of the tube.

The phase $\arg f_\alpha$ decreases monotonically as r/r_0 goes from 1 to 0. The phase delay reflects the fact that the heat exchange process needs finite time before the equilibrium state is achieved. When $\omega\tau_\alpha = 0.1$, the phase delay becomes nearly zero, meaning that the time for the heat exchange process is sufficiently smaller than the oscillation period. On the other hand, a remarkable phase delay is visible for $1 < \omega\tau_\alpha$. As a result, the gas is always subjected to the lateral thermal conduction, and therefore the gas motion becomes thermodynamically irreversible.

Figures 5.4(a)–(d) illustrate more directly the temporal entropy change of the gas parcel, where the snap shot is shown at a constant time interval. We see that the entropy change takes place only near the wall surface with larger $\omega\tau_\alpha$, whereas the entropy change occurs fully in the entire cross-section with smaller $\omega\tau_\alpha$.

5.5.2   *Cross-sectional average of entropy fluctuation*

The entropy fluctuation depends on r through the r dependence of f_α. By using $\chi_\alpha = \langle f_\alpha \rangle_r$, the cross-sectional average of the entropy fluctuation is expressed as

$$\langle S_1 \rangle_r = \chi_\alpha S_1(r_0),\qquad(5.23)$$

where $S_1(r_0) = (\partial S/\partial p)_T p_1$. The function χ_α is given by using the first-order Bessel function J_1 as

$$\chi_\alpha = \frac{2J_1(\eta_{\alpha 0})}{\eta_{\alpha 0} J_0(\eta_{\alpha 0})}.\qquad(5.24)$$

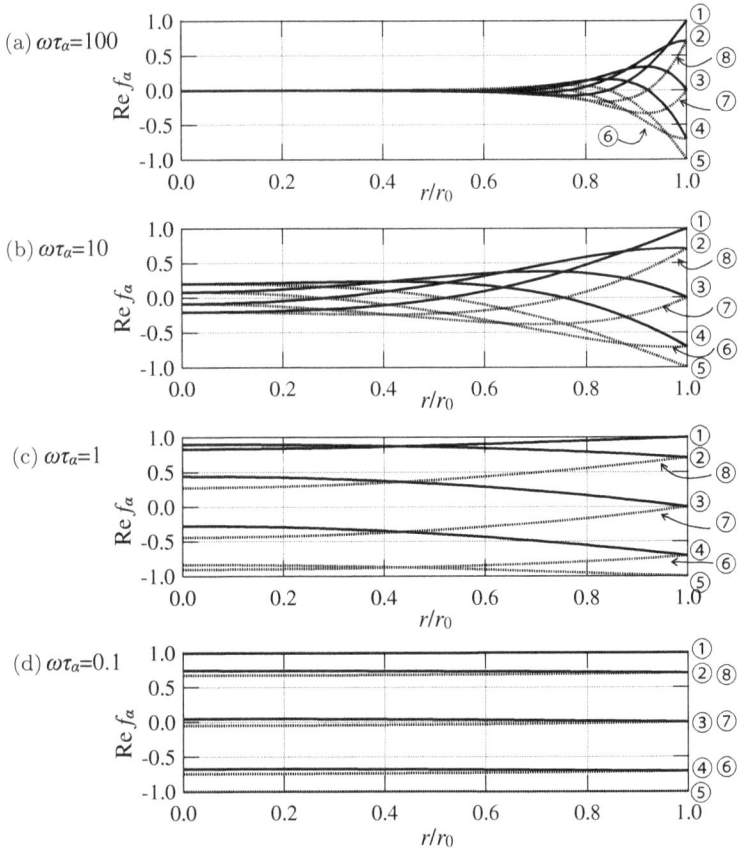

Figure 5.4: Temporal radial profile of entropy oscillation of the gas. The gas entropy changes with time in order of ①, ②, ③, ④, ⑤, ⑥, ⑦, ⑧.

Because f_α is a complex quantity, χ_α is also a complex quantity:

$$\chi_\alpha = \chi'_\alpha + i\chi''_\alpha. \tag{5.25}$$

The real part χ'_α represents a measure of reversible heat exchange process, whereas the imaginary part χ''_α gives a measure of irreversible heat exchange process.

Figure 5.5 presents the phasors representing the isothermal entropy oscillation $S_1(r_0)$ and the cross-sectional averaged entropy

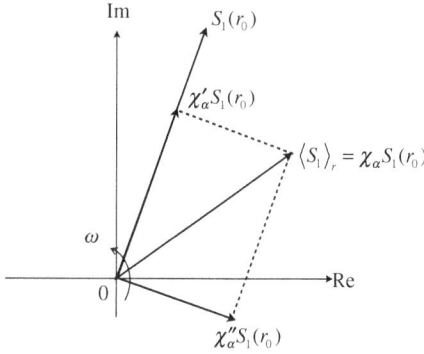

Figure 5.5: Phasor representation of the isothermal entropy oscillation $S_1(r_0)$ and the cross-sectional averaged entropy oscillation $\langle S_1 \rangle_r$.

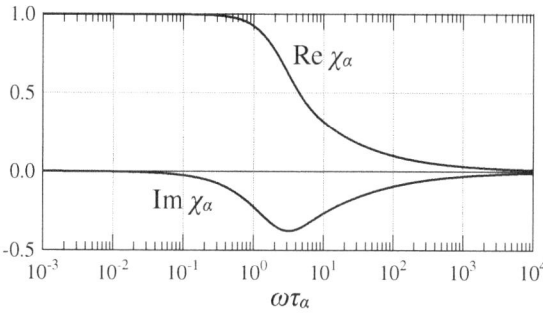

Figure 5.6: $\omega\tau_\alpha$ dependence of the real part χ'_α and the imaginary part χ''_α.

oscillation $\langle S_1 \rangle_r$. Also, Fig. 5.6 shows $\omega\tau_\alpha$ dependence of the real part χ'_α and the imaginary part χ''_α. In the isothermal limit of $\omega\tau_\alpha \ll 1$, $\chi_\alpha \sim 1$ is satisfied, because $\chi'_\alpha \sim 1$ and $\chi''_\alpha \sim 0$. On the other hand, in the adiabatic limit of $\omega\tau_\alpha \gg 1$, $\chi_\alpha \sim 0$ since both χ'_α and χ''_α are close to zero. In the intermediate region, $|\chi''_\alpha|$ shows a maximum. Since χ''_α is always negative, $\langle S_1 \rangle_r$ delays in time relative to the isothermal entropy $S_1(r_0)$, as shown by phasors in Fig. 5.5. The time delay becomes most remarkable when $\omega\tau_\alpha \sim 3$. In other words, the irreversibility of heat exchange process should be taken into account with $\omega\tau_\alpha$ around 3.

5.6 Entropy Fluctuation in Tubes with Temperature Gradient

5.6.1 *Inviscid fluid*

For an inviscid fluid with $\nu = 0$, Eq. (5.16) is reduced to

$$S_1 = \frac{\alpha}{i\omega}\Delta_\perp S_1 - \left(\frac{\partial S}{\partial T}\right)_p \frac{dT_m}{dx}\frac{u_1}{i\omega}. \qquad (5.26)$$

Suppose that the solution is expressed as

$$S_1 = f_\alpha \left(\frac{\partial S}{\partial p}\right)_T p_1 + (C_1 f_\alpha + C_2)\left(\frac{\partial S}{\partial T}\right)_p \frac{dT_m}{dx}\frac{u_1}{i\omega},$$

where C_1 and C_2 are unknown constants. Inserting into Eq. (5.26) yields

$$C_2 = -1,$$

because u_1 is independent of r for an inviscid fluid, and also f_α satisfies the relation $f_\alpha = [\alpha/(i\omega)]\Delta_\perp f_\alpha$. In order to meet the boundary condition given in Eq. (5.17), C_1 should be

$$C_1 = 1.$$

Therefore, the Eulerian entropy fluctuation for an inviscid fluid is represented by

$$S_1 = f_\alpha \left(\frac{\partial S}{\partial p}\right)_T p_1 + (f_\alpha - 1)\left(\frac{\partial S}{\partial T}\right)_p \frac{dT_m}{dx}\frac{u_1}{i\omega}. \qquad (5.27)$$

The Lagrangian entropy fluctuation is simply given by adding the entropy fluctuation caused by the displacement $\xi_1 = u_1/(i\omega)$

$$\frac{dS_m}{dx}\xi_1 = \left(\frac{\partial S}{\partial T}\right)_p \frac{dT_m}{dx}\xi_1.$$

If we introduce

$$S_{1L} = \left(\frac{\partial S}{\partial p}\right)_T p_1 + \left(\frac{\partial S}{\partial T}\right)_p \frac{dT_m}{dx}\xi_1, \qquad (5.28)$$

the Lagrangian entropy fluctuation are expressed as

$$S_1 = f_\alpha S_{1L}. \qquad (5.29)$$

The first term on the right hand side of Eq. (5.28), $(\partial S/\partial p)_T p_1$, represents the entropy oscillation associated with the isothermal pressure oscillation. In most fluids, $(\partial S/\partial p)_T$ is negative. Therefore, this component of entropy oscillation is in anti-phase with pressure oscillation. In other words, when pressure rises, entropy decreases. Decrease of entropy is brought about by the heat transfer to the channel wall from the fluid. In contrast, when the entropy increases, the fluid absorbs heat from the channel wall.

The second term of Eq. (5.28), $(\partial S/\partial T)_p(dT_m/dx)\xi_1$, originates from the displacement along the channel with temperature gradient, as shown in Fig. 5.7. Since $(\partial S/\partial T)_p$ is positive, the entropy rises as the fluid moves in a positive direction when the temperature gradient is positive. Thus, this component of entropy oscillation is temporally in phase with the displacement oscillation. Again, the origin of entropy change is the heat exchange between the fluid and the channel wall.

The factor f_α in Eq. (5.29) represents thermal relaxation over the cross-section. The cross-sectional average of the Lagrangian entropy

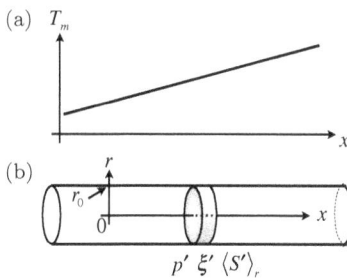

Figure 5.7: Fluid parcel in a flow channel (a) and temperature distribution along the channel (b).

oscillation is given by

$$\langle S_1 \rangle_r = \chi_\alpha S_{1L}. \tag{5.30}$$

5.6.2 *Viscous fluid*

To find a solution of Eq. (5.16) in the case of viscous fluid, we assume the Eulerian entropy fluctuation S_1 of the form

$$S_1 = \left(\frac{\partial S}{\partial p}\right)_T f_\alpha p_1 + \left(\frac{\partial S}{\partial T}\right)_p \frac{dT_m}{dx} \frac{u_{1r}}{i\omega} c \tag{5.31}$$

where c is an unknown factor to be determined. Because $S_1(r_0) = (\partial S/\partial p)_T p_1$ when $r = r_0$, the relation

$$c(r_0) = 0 \tag{5.32}$$

must hold. By inserting Eq. (5.31) into Eq. (5.16) and by using the relation $f_\alpha = \alpha \Delta_\perp f_\alpha / (i\omega)$, we have the equation that c should satisfy:

$$c = \frac{\alpha}{i\omega} \Delta_\perp c - \frac{1 - f_\nu}{1 - \chi_\nu}. \tag{5.33}$$

Here, c is assumed as

$$c = C_1 (1 - f_\alpha) + C_2 (1 - f_\nu).$$

Note that the boundary condition Eq. (5.32) is automatically satisfied. By inserting the assumed solution into Eq. (5.33), we are able to have

$$C_1 + C_2 + \frac{1}{1 - \chi_\nu} = f_\nu \left[\left(1 - \frac{\alpha}{\nu}\right) C_2 + \frac{1}{1 - \chi_\nu}\right].$$

The left hand side of the equation above is independent of r, whereas the right hand side depends on r through f_ν. Therefore,

$$C_1 + C_2 + \frac{1}{1 - \chi_\nu} = 0, \quad \left(1 - \frac{\alpha}{\nu}\right) C_2 + \frac{1}{1 - \chi_\nu} = 0.$$

The coupled equations with respect to C_1 and C_2 obtained above can be solved as

$$C_1 = -\frac{1}{(1 - \chi_\nu)(1 - \sigma)}, \quad C_2 = \frac{\sigma}{(1 - \chi_\nu)(1 - \sigma)},$$

where

$$\sigma = \frac{\nu}{\alpha}$$

denotes the Prandtl number of the fluid. Finally, we are able to find that c in Eq. (5.31) is given by

$$c = \frac{(f_\alpha - 1) - \sigma(f_\nu - 1)}{(1 - \chi_\nu)(1 - \sigma)}. \tag{5.34}$$

The complex amplitude of the cross-sectional mean of the Eulerian entropy fluctuation is expressed as

$$\langle S_1 \rangle_r = \left(\frac{\partial S}{\partial p}\right)_T \chi_\alpha p_1 + \left(\frac{\partial S}{\partial T}\right)_p \frac{dT_m}{dx} \langle c \rangle_r \frac{u_{1r}}{i\omega}. \tag{5.35}$$

The Lagrangian entropy fluctuation is gained from the Eulerian one by using the result of Section 5.2 as

$$S_1 = \left(\frac{\partial S}{\partial p}\right)_T f_\alpha p_1 + \left(\frac{\partial S}{\partial T}\right)_p \frac{dT_m}{dx} b\xi_{1r} \tag{5.36}$$

where

$$b = \frac{f_\alpha - f_\nu}{(1 - \chi_\nu)(1 - \sigma)}. \tag{5.37}$$

Note that we have used $\xi_{1r} = u_{1r}/(i\omega)$. The cross-sectional mean of the Lagrangian entropy fluctuation is given by

$$\langle S_1 \rangle_r = \left(\frac{\partial S}{\partial p}\right)_T \chi_\alpha p_1 + \left(\frac{\partial S}{\partial T}\right)_p \frac{dT_m}{dx} \langle b \rangle_r \xi_{1r} \tag{5.38}$$

where

$$\langle b \rangle_r = \frac{\chi_\alpha - \chi_\nu}{(1 - \chi_\nu)(1 - \sigma)}. \tag{5.39}$$

One can see that the following relations hold between b and c:

$$b = c + \frac{1 - f_\nu}{1 - \chi_\nu}, \tag{5.40}$$

$$\langle b \rangle_r = \langle c \rangle_r + 1. \tag{5.41}$$

5.7 Fluctuations of Temperature and Density

We have evaluated the entropy fluctuation S' of the gas oscillating in a tube. Let's use it to derive temperature fluctuation T' and density fluctuation ρ' of the thermoviscous fluid.

5.7.1 *Temperature fluctuation*

Temperature fluctuation T' can be expressed as $T' = (\partial T / \partial p)_S p' + (\partial T / \partial S)_p S'$. Therefore, the corresponding complex amplitudes satisfy the following equation:

$$T_1 = \left(\frac{\partial T}{\partial p} \right)_S p_1 + \left(\frac{\partial T}{\partial S} \right)_p S_1.$$

Inserting the Eulerian entropy S_1 in Eq. (5.31) yields

$$T_1 = \left(\frac{\partial T}{\partial p} \right)_S p_1 + \left(\frac{\partial T}{\partial S} \right)_p \left[\left(\frac{\partial S}{\partial p} \right)_T f_\alpha p_1 + \left(\frac{\partial S}{\partial T} \right)_p \frac{dT_m}{dx} c \frac{u_{1r}}{i\omega} \right].$$

By using a formula of partial derivative, we see

$$\left(\frac{\partial T}{\partial S} \right)_p \left(\frac{\partial S}{\partial p} \right)_T = - \left(\frac{\partial T}{\partial p} \right)_S.$$

Therefore, the Eulerian temperature T_1 is reduced to

$$T_1 = \left(\frac{\partial T}{\partial p} \right)_S (1 - f_\alpha) p_1 + \frac{dT_m}{dx} c \frac{u_{1r}}{i\omega}. \tag{5.42}$$

Then, the Lagrangian temperature is given by

$$T_1 = \left(\frac{\partial T}{\partial p} \right)_S (1 - f_\alpha) p_1 + \frac{dT_m}{dx} b \frac{u_{1r}}{i\omega}. \tag{5.43}$$

In the absence of the temperature gradient $(dT_m/dx = 0)$, the temperature fluctuation is expressed by

$$T_1 = \left(\frac{\partial T}{\partial p}\right)_S (1 - f_\alpha)p_1.$$

Hence, the r dependence of T_1 is governed by the factor $1 - f_\alpha$. Since the velocity of the viscous fluid is expressed by

$$u_1 = \frac{1 - f_\nu}{1 - \chi_\nu} u_{1r},$$

r dependence of T_1 is the same as that of u_1 if the Prandtl number σ is $\sigma = 1$. When $\sigma < 1$, the interaction with the wall is more remarkable in T_1 than in u_1. Figure 5.8 compares the factor $1 - f_\alpha$ and $1 - f_\nu$ when $\sigma = 0.67$. It is depicted that the thermal boundary layer thickness is greater than the viscous boundary layer thickness.

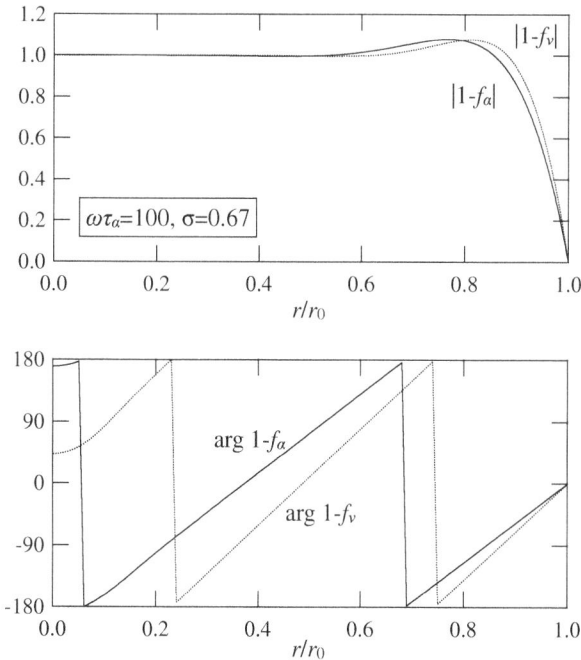

Figure 5.8: Comparison between $1 - f_\alpha$ and $1 - f_\nu$ when $\omega\tau_\alpha = 100$ and $\sigma = 0.67$.

5.7.2 *Density fluctuation*

Consider here the density fluctuation ρ' and its complex amplitude ρ_1 based on the temperature fluctuation in Eq. (5.42). The density fluctuation ρ' can be expressed as $\rho' = (\partial\rho/\partial p)_T p' + (\partial\rho/\partial T)_p T'$. Therefore, the associated complex amplitudes satisfy the following equation:

$$\rho_1 = \left(\frac{\partial\rho}{\partial p}\right)_T p_1 + \left(\frac{\partial\rho}{\partial T}\right)_p T_1.$$

By inserting T_1 in Eq. (5.42), we have

$$\rho_1 = \left(\frac{\partial\rho}{\partial p}\right)_T p_1 + \left(\frac{\partial\rho}{\partial T}\right)_p \left[\left(\frac{\partial T}{\partial p}\right)_S (1 - f_\alpha)p_1 + \frac{dT_m}{dx}c\frac{u_{1r}}{i\omega}\right].$$

$$(5.44)$$

By using a mathematical formula of partial derivative

$$\left(\frac{\partial\rho}{\partial T}\right)_p\left(\frac{\partial T}{\partial p}\right)_S = \left(\frac{\partial\rho}{\partial p}\right)_S - \left(\frac{\partial\rho}{\partial p}\right)_T$$

and also by using the thermodynamic quantities

$$\text{isothermal compressibility: } K_T = \frac{1}{\rho_m}\left(\frac{\partial\rho}{\partial p}\right)_T$$

$$\text{adiabatic compressibility: } K_S = \frac{1}{\rho_m}\left(\frac{\partial\rho}{\partial p}\right)_S$$

$$\text{thermal expansion coefficient : } \beta = -\frac{1}{\rho_m}\left(\frac{\partial\rho}{\partial T}\right)_p$$

the density amplitude ρ_1 is reduced to

$$\rho_1 = \rho_m\left[K_S + (K_T - K_S)f_\alpha\right]p_1 - \rho_m\beta\frac{dT_m}{dx}c\frac{u_{1r}}{i\omega}. \qquad (5.45)$$

Therefore, the cross-sectional mean density $\langle \rho_1 \rangle_r$ is given by

$$\langle \rho_1 \rangle_r = \rho_m K_E p_1 - \rho_m \beta \frac{dT_m}{dx} \langle c \rangle_r \frac{u_{1r}}{i\omega}, \qquad (5.46)$$

where K_E is defined by

$$K_E = K_S + (K_T - K_S)\chi_\alpha, \qquad (5.47)$$

which means the effective compressibility of the gas oscillating in a tube.

The Lagrangian density is given by

$$\rho_1 = \rho_m \left[K_S + (K_T - K_S)f_\alpha \right] p_1 - \rho_m \beta \frac{dT_m}{dx} b\xi_{1r}. \qquad (5.48)$$

Correspondingly, the cross-sectional mean of the Lagrangian density is given by

$$\langle \rho_1 \rangle_r = \rho_m K_E p_1 - \rho_m \beta_E \frac{dT_m}{dx} \xi_{1r}, \qquad (5.49)$$

where β_E indicates the effective thermal expansion coefficient of the gas and is given by

$$\beta_E = \beta \langle b \rangle_r. \qquad (5.50)$$

If we take the cross-sectional average of the continuity equation in Eq. (5.11), we have

$$\frac{\partial \langle u' \rangle_r}{\partial x} = -\frac{1}{\rho_m} \frac{d\langle \rho' \rangle_r}{dt}. \qquad (5.51)$$

By introducing the complex amplitude and using Eq. (5.49) to rewrite Eq. (5.51), we obtain

$$\frac{du_{1r}}{dx} = -i\omega K_E p_1 + \beta_E \frac{dT_m}{dx} u_{1r}. \qquad (5.52)$$

This equation represents the continuity equation expressed by using the solution of general equation of heat transfer of the thermoviscous fluid oscillating in a tube.

5.8   Wave Equation of Sound Waves in a Tube

By inserting the momentum equation expressed by the cross-sectional mean velocity

$$u_{1r} = \frac{i}{\omega \rho_m} \frac{dp_1}{dx} (1 - \chi_\nu)$$

into the continuity equation in Eq. (5.52) gives

$$K_E p_1 - \frac{\beta_E}{\omega^2 \rho_m} \frac{dT_m}{dx} \frac{dp_1}{dx} (1 - \chi_\nu) + \frac{1}{\omega^2} \frac{d}{dx} \left[\frac{1 - \chi_\nu}{\rho_m} \frac{dp_1}{dx} \right] = 0, \quad (5.53)$$

which serves as the wave equation of sound waves in a tube. When the tube has a uniform temperature, the wave equation is represented by

$$p_1 + \frac{1 - \chi_\nu}{\omega^2 \rho_m K_E} \frac{d^2 p_1}{dx^2} = 0. \quad (5.54)$$

Assume the solution of the form

$$p_1 \propto e^{-ikx} = e^{\mathrm{Im}[k]x} e^{-i\mathrm{Re}[k]x}$$

where $k = \mathrm{Re}[k] + i\mathrm{Im}[k]$ is a complex wavenumber of sound waves propagating in the uniform temperature tube. It should be noted that the sound speed is given by $\omega/\mathrm{Re}[k]$, whereas the attenuation constant is given by $-\mathrm{Im}[k]$. Inserting $p_1 = e^{-ikx}$ into the wave equation gives

$$k^2 = \frac{\omega^2 \rho_m K_E}{1 - \chi_\nu} = \omega^2 \rho_m \frac{K_S + (K_T - K_S)\chi_\alpha}{1 - \chi_\nu}, \quad (5.55)$$

which can be rewritten as

$$\frac{k}{k_S} = \sqrt{\frac{1 + (K_T/K_S - 1)\chi_\alpha}{1 - \chi_\nu}}, \quad (5.56)$$

by using the symbols:

$$c_S = \sqrt{\left(\frac{\partial p}{\partial \rho}\right)_S} = \sqrt{\frac{1}{\rho_m K_S}}, \quad k_S = \frac{\omega}{c_S}.$$

Because $K_T/K_S = \gamma$ represents the ratio of specific heat at constant pressure to specific heat at constant volume, the wave number is simply expressed as

$$\frac{k}{k_S} = \sqrt{\frac{1 + (\gamma - 1)\chi_\alpha}{1 - \chi_\nu}}. \tag{5.57}$$

Therefore, the propagation constant $\Gamma = ik/k_S$ introduced in Chapter 2 is given by

$$\Gamma = \sqrt{-\frac{1 + (\gamma - 1)\chi_\alpha}{1 - \chi_\nu}}.$$

Asymptotic form of k introduced in Chapter 2 is obtained from Eq. (5.57) as follows. The limiting values of χ_α and χ_ν are expressed by

$$\chi_j = \begin{cases} 1 - \dfrac{(\omega\tau_j)^2}{12} - i\dfrac{\omega\tau_j}{4} & (\omega\tau_j \ll \pi) \\[3mm] \sqrt{\dfrac{2}{\omega\tau_j}}e^{-i\pi/4} & (\omega\tau_j \gg \pi) \end{cases} \quad (j = \alpha,\ \nu),$$

which is already shown in Chapter 3.

Therefore, for $\omega\tau_j \ll \pi$, we have

$$\frac{k}{k_S} = \sqrt{\frac{1 + (\gamma - 1)\left[1 - \dfrac{(\omega\tau_\alpha)^2}{12} - i\dfrac{\omega\tau_\alpha}{4}\right]}{1 - \left[1 - \dfrac{(\omega\tau_\nu)^2}{12} - i\dfrac{\omega\tau_\nu}{4}\right]}}.$$

Because $\omega\tau_j \ll \pi$, the expression above is simplified as

$$\frac{k}{k_S} \sim \sqrt{\frac{\gamma}{i\omega\tau_\nu/4}} = (1 - i)\sqrt{\frac{2\gamma\sigma}{\omega\tau_\alpha}}.$$

On the other hand, for $\omega \tau_j \gg \pi$, we have

$$\frac{k}{k_S} = \sqrt{\frac{1}{1 - \chi_\nu}}\sqrt{1 + (\gamma - 1)\chi_\alpha}$$

$$\sim \left(1 + \frac{\chi_\nu}{2}\right)\left(1 + \frac{(\gamma - 1)\chi_\alpha}{2}\right)$$

$$\sim 1 + \frac{1}{2}\left[(\gamma - 1)\chi_\alpha + \chi_\nu\right].$$

Because $\omega \tau_j \gg \pi$, the expression above is simplified by using the asymptotic form of χ_j $(j = \alpha,\ \nu)$ as follows:

$$\frac{k}{k_S} \sim 1 + \frac{1}{2}\left[(\gamma - 1)\sqrt{\frac{2}{\omega\tau_\alpha}}e^{-i\pi/4} + \sqrt{\frac{2}{\omega\tau_\nu}}e^{-i\pi/4}\right]$$

$$= 1 + \frac{(1 - i)(\gamma - 1 + \sqrt{\sigma})}{2\sqrt{\omega\tau_\alpha}}.$$

5.9 Appendix: Basic Equations of Hydrodynamics

5.9.1 *Conservation laws*

We derive here the basic equations of hydrodynamics based on the conservation laws of mass, momentum, and energy. We employ the cartesian coordinates $\boldsymbol{x} = (x, y, z) = (x_1, x_2, x_3)$ and time t to express the velocity field $\boldsymbol{u}(\boldsymbol{x}, t) = (u(\boldsymbol{x}, t), v(\boldsymbol{x}, t), z(\boldsymbol{x}, t)) = (u_1(\boldsymbol{x}, t), u_2(\boldsymbol{x}, t), u_3(\boldsymbol{x}, t))$, pressure field $p(\boldsymbol{x}, t)$, and density field $\rho_m(\boldsymbol{x}, t)$.

Suppose that a volume V enclosed by a surface S is fixed in a space. The conservation law for the quantities contained in V is expressed as

$$\frac{\partial}{\partial t}\iiint_V [\cdots]\, dV + \iint_S (\cdots)_n\, dS = 0,$$

where $[\cdots]$ includes the state quantities such as density, momentum, and energy per unit volume, $(\cdots)$ represents the corresponding flux density such as mass flow, momentum flow, and energy flow per unit cross-sectional area. The subscript n means to take the vector component along the outward normal direction of the surface element dS. Therefore, $(\cdots)_n \, dS$ expresses the flow rate going out of V through dS. The first term represents the increasing rate of the quantity contained in V, whereas the second term represents the flow rate going out of V. Because the sum of them is equal to zero, the equation means the conservation law.

By using Gauss' theorem, surface integral can be expressed by volume integral. Suppose that the outward normal vector of the surface S that encloses the volume V is given by $\boldsymbol{n} = (n_1, n_2, n_3)$. Gauss' theorem for a vector $\boldsymbol{A} = (A_1, A_2, A_3)$ reads

$$\iint_S \boldsymbol{A} \cdot \boldsymbol{n} \, dS = \iiint_V \operatorname{div} \boldsymbol{A} \, dV,$$

where $\operatorname{div} \boldsymbol{A}$ is given by

$$\operatorname{div} \boldsymbol{A} = \frac{\partial A_1}{\partial x_1} + \frac{\partial A_2}{\partial x_2} + \frac{\partial A_3}{\partial x_3}.$$

Therefore, the conservation law can be expressed in a differential form as

$$\frac{\partial}{\partial t}[\cdots] + \frac{\partial}{\partial x_k}(\cdots)_k = 0,$$

where the second term in the left hand side represents the abbreviation of

$$\sum_{k=1}^{3} \frac{\partial}{\partial x_k}(\cdots)_k.$$

5.9.2 *Continuity equation*

The continuity equation for the volume V enclosed by the surface S in Fig. 5.9 is expressed as

$$\frac{\partial}{\partial t} \iiint_V \rho \, dV + \iint_S \rho u_n \, dS = 0,$$

where $u_n = \boldsymbol{u} \cdot \boldsymbol{n}$ means the vector component of the velocity $\boldsymbol{u}$ in the direction of $\boldsymbol{n}$. The first term on the left hand side of the equation represents the increasing rate of mass contained in V per

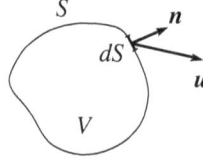

Figure 5.9: Volume V, surface S, outward normal vector $\boldsymbol{n}$, and velocity $\boldsymbol{u}$.

unit time, whereas the second term represents the flow rate going out of V through S per unit time. By using Gauss' theorem, the following equation is obtained

$$\frac{\partial \rho}{\partial t} + \operatorname{div}(\rho \boldsymbol{u}) = 0,$$

which is called the continuity equation.

5.9.3   Equation of motion (Navier-Stokes equation)

The conservation of momentum in a volume V enclosed by a surface S (see Fig. 5.10) is given by

$$\frac{\partial}{\partial t} \iiint_V \rho \boldsymbol{u} \, dV + \iint_S \rho \boldsymbol{u} u_n \, dS + \iint_S (-\boldsymbol{p}_n) \, dS = 0$$

when a body force like gravitational force is absent. Here, $\boldsymbol{p}_n$ represents the stress vector that points outward at the surface S, which can be expressed as $\boldsymbol{p}_n = \boldsymbol{P} \cdot \boldsymbol{n} = p_{ij} n_j$ by using stress tensor $\boldsymbol{P} = \{p_{ij}\}$. The first term on the left hand side represents the increase rate of the momentum in V per unit time, the second term the flow rate going out of V per unit time due to the mass flow, and the third the contribution by the stress. The sum of the second and the third terms, $\rho \boldsymbol{u} u_n - \boldsymbol{p}_n$, gives the flow rate of the momentum that leaves V through S per unit time.

The i-th component of the equation reads

$$\frac{\partial}{\partial t} \iiint_V \rho u_i \, dV + \iint_S (\rho u_i u_j n_j - p_{ij} n_j) dS = 0.$$

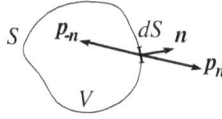

Figure 5.10:   Volume V, surface S, outward normal vector $\boldsymbol{n}$, and stress vector $\boldsymbol{p}_n$.

From Gauss' theorem, we have

$$\iint_S \rho u_i u_j n_j \, dS = \iiint_V \frac{\partial}{\partial x_j} (\rho u_i u_j) \, dV,$$

and

$$\iint_S p_{ij} n_j \, dS = \iiint_V \frac{\partial p_{ij}}{\partial x_j} \, dV.$$

Hence, the following equation is obtained:

$$\frac{\partial}{\partial t}(\rho u_i) + \frac{\partial}{\partial x_j}(\rho u_i u_j) - \frac{\partial p_{ij}}{\partial x_j} = 0.$$

By combining with the continuity equation

$$\frac{\partial \rho}{\partial t} + u_j \frac{\partial \rho}{\partial x_j} + \rho \frac{\partial u_j}{\partial x_j} = 0, \tag{5.58}$$

the equation is expressed as

$$\rho \left(\frac{\partial u_i}{\partial t} + u_j \frac{\partial u_i}{\partial x_j} \right) - \frac{\partial p_{ij}}{\partial x_j} = 0. \tag{5.59}$$

For a Newtonian fluid, the stress tensor is expressed as

$$p_{ij} = -p \delta_{ij} + \lambda (\mathrm{div}\, \boldsymbol{u}) \delta_{ij} + 2\mu e_{ij}$$

where δ_{ij} denotes Cronecker's delta; $\delta_{ij} = 1$ when $i = j$, and $\delta_{ij} = 0$ otherwise. Also, mu means the viscosity; λ is the second viscosity, and e_{ij} is the strain rate tensor

$$e_{ij} = \frac{1}{2} \left(\frac{\partial u_i}{\partial x_j} + \frac{\partial u_j}{\partial x_i} \right).$$

If we write $D = \text{div } \boldsymbol{u}$, the stress tensor included in Eq. (5.59) is explicitly given by

$$\frac{\partial p_{ij}}{\partial x_j} = -\delta_{ij}\frac{\partial p}{\partial x_j} + \lambda\delta_{ij}\frac{\partial D}{\partial x_j} + \mu\frac{\partial}{\partial x_j}\left(\frac{\partial u_i}{\partial x_j} + \frac{\partial u_j}{\partial x_i}\right),$$

where

$$\delta_{ij}\frac{\partial}{\partial x_j} = \frac{\partial}{\partial x_i},$$

$$\frac{\partial^2}{\partial x_j \partial x_j} = \frac{\partial^2}{\partial x^2} + \frac{\partial^2}{\partial y^2} + \frac{\partial^2}{\partial z^2}(= \Delta),$$

and

$$\frac{\partial}{\partial x_j}\left(\frac{\partial u_j}{\partial x_i}\right) = \frac{\partial}{\partial x_i}\left(\frac{\partial u_j}{\partial x_j}\right) = \frac{\partial D}{\partial x_i}.$$

Therefore,

$$\frac{\partial p_{ij}}{\partial x_j} = \frac{\partial p}{\partial x_i} + (\lambda + \mu)\frac{\partial D}{\partial x_i} + \mu\Delta\boldsymbol{u} = -\nabla p + (\lambda + \mu)\nabla D + \mu\Delta\boldsymbol{u}.$$

Finally, we are able to arrive at the Navier-Stokes equation

$$\rho\left(\frac{\partial \boldsymbol{u}}{\partial t} + (\boldsymbol{u}\cdot\nabla)\boldsymbol{u}\right) = -\nabla p + \mu\Delta\boldsymbol{u} + (\lambda + \mu)\nabla D.$$

In the case when compression is negligibly small ($D = 0$), we have

$$\rho\left(\frac{\partial \boldsymbol{u}}{\partial t} + (\boldsymbol{u}\cdot\nabla)\boldsymbol{u}\right) = -\nabla p + \mu\Delta\boldsymbol{u}.$$

This equation is written in a short form by using the Lagrangian derivative as

$$\rho\frac{d\boldsymbol{u}}{dt} = -\nabla p + \mu\Delta\boldsymbol{u}. \tag{5.60}$$

5.9.4   *Energy equation*

Consider energy conservation of a fluid in a volume V enclosed by a surface S. Suppose U expresses the specific internal energy of the fluid, and κ denotes the thermal conductivity. The increasing rate of the energy per unit time is represented by

$$\frac{\partial}{\partial t} \iiint_V \rho \left(\frac{1}{2}|u|^2 + U\right) dV.$$

The flow rate of energy per unit time that leaves V due to mass flow is expressed as

$$\iint_S \rho \left(\frac{1}{2}|u|^2 + U\right) u_n \, dS,$$

whereas the energy flow rate due to thermal conduction is given by

$$\iint_S \boldsymbol{\theta} \cdot \boldsymbol{n} \, dS,$$

with $\boldsymbol{\theta} = -\kappa \nabla T$. Also, the work done by the neighboring fluid through surface per unit time is given by

$$\iint_S -p_n \cdot u \, dS = - \iint_S (\boldsymbol{P} \cdot \boldsymbol{n}) \cdot u \, dS.$$

Therefore, energy conservation is represented by the following equation:

$$\frac{\partial}{\partial t} \iiint_V \rho \left(\frac{1}{2}|u|^2 + U\right) dV$$

$$+ \iint_S \rho \left(\frac{1}{2}|u|^2 + U\right) u_n \, dS + \iint_S \boldsymbol{\theta} \cdot \boldsymbol{n} \, dS - \iint_S (\boldsymbol{P} \cdot \boldsymbol{n}) \cdot u \, dS = 0.$$

Use of Gauss' theorem leads to

$$\iint_S \theta_j n_j \, dS = \iiint_V \frac{\partial \theta_j}{\partial x_j} \, dV,$$

and

$$\iint_S p_{ij} n_j u_i \, dS = \iiint_V \frac{\partial}{\partial x_j} (p_{ij} u_i) \, dV.$$

Therefore,

$$\frac{\partial}{\partial t}\left[\rho\left(\frac{1}{2}|\boldsymbol{u}|^2 + U\right)\right] + \frac{\partial}{\partial x_j}\left[-p_{ij}u_i + \rho\left(\frac{1}{2}|\boldsymbol{u}|^2 + U\right)u_j + \theta_j\right] = 0.$$

(5.61)

This equation is called the energy equation.

Let us express the stress tensor for a Newtonian fluid, $p_{ij} = -p\delta_{ij} + \lambda D\delta_{ij} + 2\mu e_{ij}$, by $p_{ij} = -p\delta_{ij} + \sum_{ij}$ for brevity. If we use the specific enthalpy $H = U + p/\rho$, the energy equation is rewritten as

$$\frac{\partial}{\partial t}\left[\rho\left(\frac{1}{2}|\boldsymbol{u}|^2 + U\right)\right] + \frac{\partial}{\partial x_j}\left[\rho H u_j + \frac{1}{2}|\boldsymbol{u}|^2 u_j - \sum_{ij} u_i + \theta_j\right] = 0.$$

The energy flow is in the second term of the left hand side of the equation:

$$\rho H\boldsymbol{u} + \frac{1}{2}|\boldsymbol{u}|^2\boldsymbol{u} - \boldsymbol{u}\cdot\sum + \boldsymbol{\theta}.$$

In a flow field with relatively small velocity, the first term dominates over the second and third terms. Therefore, the enthalpy flux density $\rho H\boldsymbol{u}$ serves as the energy flux density due to the fluid motion.

The first term on the left hand side of Eq. (5.61) is transformed by using the continuity equation in Eq. (5.58) as

$$\frac{\partial}{\partial t}\left[\rho\left(\frac{1}{2}|\boldsymbol{u}|^2 + U\right)\right]$$

$$= \rho\frac{\partial}{\partial t}\left(\frac{1}{2}|\boldsymbol{u}|^2 + U\right) + \rho u_j\frac{\partial}{\partial x_j}\left(\frac{1}{2}|\boldsymbol{u}|^2 + U\right)$$

$$- \frac{\partial}{\partial x_j}\left[\rho\left(\frac{1}{2}|\boldsymbol{u}|^2 + U\right)u_j\right]$$

$$= \rho\frac{d}{dt}\left(\frac{1}{2}|\boldsymbol{u}|^2 + U\right) - \frac{\partial}{\partial x_j}\left[\rho\left(\frac{1}{2}|\boldsymbol{u}|^2 + U\right)u_j\right].$$

Therefore, Eq. (5.61) is simplified as

$$\frac{d}{dt}\left(\frac{1}{2}|\boldsymbol{u}|^2 + U\right) = \frac{1}{\rho}\frac{\partial}{\partial x_j}(p_{ij}u_i - \theta_j).$$

(5.62)

Furthermore, we notice from Eq. (5.59) that

$$\frac{d}{dt}\left(\frac{1}{2}|\boldsymbol{u}|^2\right) = \frac{1}{\rho}\frac{\partial p_{ij}}{\partial x_j}u_i \tag{5.63}$$

should hold. From these two equations, we have

$$\frac{dU}{dt} = \frac{1}{\rho}p_{ij}\frac{\partial u_i}{\partial x_j} - \frac{1}{\rho}\frac{\partial \theta_j}{\partial x_j}.$$

Since the stress tensor of a Newtonian fluid is $p_{ij} = -p\delta_{ij} + \lambda D\delta_{ij} + 2\mu e_{ij}$, it is found that

$$p_{ij}\frac{\partial u_i}{\partial x_j} = -pD + \lambda D\delta_{ij}\frac{\partial u_i}{\partial x_j} + 2\mu e_{ij}\frac{\partial u_i}{\partial x_j} = -pD + \lambda D^2 + 2\mu e_{ij}^2.$$

If we define the dissipation function by

$$\Phi = \lambda D^2 + 2\mu e_{ij}^2,$$

the equation is written as

$$\frac{dU}{dt} = -\frac{p}{\rho}D + \frac{\Phi}{\rho} - \frac{1}{\rho}\frac{\partial \theta_j}{\partial x_j}.$$

From the thermodynamic relation

$$TdS = dU + pd\left(\frac{1}{\rho}\right),$$

we see

$$\frac{dU}{dt} = T\frac{dS}{dt} - p\frac{d}{dt}\left(\frac{1}{\rho}\right) = T\frac{dS}{dt} + \frac{p}{\rho^2}\frac{d\rho}{dt}.$$

Hence,

$$T\frac{dS}{dt} + \frac{p}{\rho^2}\frac{d\rho}{dt} = -\frac{p}{\rho}D + \frac{\Phi}{\rho} - \frac{1}{\rho}\frac{\partial \theta_j}{\partial x_j}.$$

The continuity equation in Eq. (5.58) is given by using $D = \text{div}\,\boldsymbol{u}$ as

$$\frac{d\rho}{dt} = -\rho D.$$

Finally, we arrive at the general equation of heat transfer:

$$T\frac{dS}{dt} = -\frac{1}{\rho}\frac{\partial \theta_j}{\partial x_j} + \frac{\Phi}{\rho}.$$

Because the thermal conduction is expressed by

$$\theta_j = -\kappa\frac{\partial T}{\partial x_j},$$

the general equation of heat transfer is expressed as

$$\rho T\left(\frac{\partial S}{\partial t} + (\boldsymbol{u} \cdot \nabla)S\right) = \kappa\nabla T + \Phi.$$

5.10   Problems

1 Compare $(u \cdot \nabla)u \sim u\partial u/\partial x$ with $\partial u/\partial t$ when $u = U\cos(\omega t - k_S x)$. Express the ratio

$$\left|\frac{u\partial u/\partial x}{\partial u/\partial t}\right|$$

by using U and the adiabatic sound speed c_S. State the condition when the non-linear term $(u \cdot \nabla)u$ is negligibly smaller than the linear term $\partial u/\partial t$.

2 Consider acoustic gas oscillations between parallel plates with spacing of $2h$ shown in Fig. 5.11, when the complex pressure amplitude is p_1 and angular frequency is ω. When the plate temperature is uniform, the complex amplitude S_1 of the gas entropy satisfies

$$S_1 = \frac{\alpha}{i\omega}\frac{d^2 S_1}{dy^2}.$$

Answer the following questions.

(1) Write the boundary conditions for S_1.
(2) Find S_1.
(3) Obtain the complex temperature amplitude T_1 of the gas.

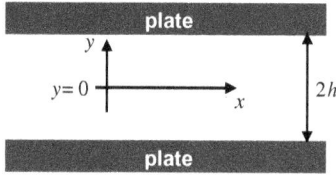

Figure 5.11: Parallel plates. y-axis is perpendicular to the plates and $y = 0$ is on the center.

Answers

1 Since $u\partial u/\partial x = k_S U^2 \cos(\omega t - k_S x)\sin(\omega t - k_S x)$, and $\partial u/\partial t = \omega U \sin(\omega t - k_S x)$, we see

$$\left|\frac{u\partial u/\partial x}{\partial u/\partial t}\right| = \left|\frac{k_S U}{\omega}\cos(\omega t - k_S x)\right| \le \left|\frac{U}{c_S}\right|.$$

Therefore the term $(u \cdot \nabla)u$ is negligibly smaller than $\partial u/\partial t$ when $U \ll c_S$.

2 (1) The boundary conditions are as follows.

$$S_1(\pm h) = \left(\frac{\partial S}{\partial p}\right)_T p_1,$$

$$\frac{dS_1(0)}{dy} = 0.$$

(2) S_1 is given by $S_1 = C_1 e^{\lambda y} + C_2 e^{-\lambda y}$, where $\lambda = \sqrt{i\omega/\alpha}$; C_1 and C_2 are unknown constants. By using the boundary conditions, we see

$$C_1 = C_2 = \frac{1}{2\cosh(\lambda h)}\left(\frac{\partial S}{\partial p}\right)_T p_1.$$

Therefore, S_1 is given by

$$S_1 = f_\alpha \left(\frac{\partial S}{\partial p}\right)_T p_1.$$

with

$$f_\alpha = \frac{\cosh(\lambda y)}{\cosh(\lambda h)}.$$

(3) Temperature is expanded by pressure and entropy as

$$T_1 = \left(\frac{\partial T}{\partial p}\right)_S p_1 + \left(\frac{\partial T}{\partial S}\right)_p S_1.$$

Therefore, we have

$$T_1 = \left(\frac{\partial T}{\partial p}\right)_S p_1 + \left(\frac{\partial T}{\partial S}\right)_p \left(\frac{\partial S}{\partial p}\right)_T f_\alpha p_1.$$

Since the relation

$$\left(\frac{\partial T}{\partial S}\right)_p \left(\frac{\partial S}{\partial p}\right)_T \left(\frac{\partial p}{\partial T}\right)_S = -1$$

holds, we obtain

$$T_1 = \left(\frac{\partial T}{\partial p}\right)_S (1 - f_\alpha) p_1.$$

Chapter 6

Components of Energy Flows and Work Source and Their Classification

Heat flux density, work flux density, and work sources are determined by pressure, displacement, and entropy oscillations of fluid parcels. Particularly for an inviscid fluid, energy flux densities and work source can be depicted by tracing these oscillations. Graphical presentation of the traces for the inviscid fluid helps to understand the physical meaning of energy flows and work source of a viscous fluid. Decomposition of pressure oscillation into the in-phase component with displacement and the 90-degree out-of-phase component with displacement provides the first step toward a comprehensive understanding of the energy flows and work source.

6.1 Work Flux Density, Heat Flux Density, and Work Source of Inviscid Fluid

As we have shown in Section 5.3, work flux density I, heat flux density Q, and work source w are formulated by using pressure oscillations p', velocity oscillations u', and entropy oscillations S' as

$$I = \langle\langle p'u' \rangle\rangle,$$

$$Q = \rho_m T_m \langle\langle S'u' \rangle\rangle$$

$$w = \rho_m \left(\frac{\partial V}{\partial S} \right)_p \left\langle\left\langle p'\frac{dS'}{dt} \right\rangle\right\rangle + W_\nu, \quad W_\nu = \langle\langle (\mu\Delta_\perp u')u' \rangle\rangle.$$

For an inviscid fluid, u' and displacement oscillation ξ' do not depend on radial coordinate r. As a result, taking the cross-sectional averages included in I, Q, and w is greatly simplified. In other words, for the inviscid fluid, I, Q, and w are reduced as follows:

$$I = \langle p'u' \rangle_t = \frac{\omega}{2\pi} \oint p' \, d\xi' \tag{6.1}$$

$$Q = \rho_m T_m \langle \langle S' \rangle_r u' \rangle_t = \frac{\omega}{2\pi} \rho_m T_m \oint \langle S' \rangle_r \, d\xi' \tag{6.2}$$

$$w = \rho_m \left(\frac{\partial V}{\partial S} \right)_p \left\langle p' \frac{d\langle S' \rangle_r}{dt} \right\rangle_t = \frac{\omega}{2\pi} \rho_m \left(\frac{\partial V}{\partial S} \right)_p \oint p' \, d\langle S' \rangle_r, \tag{6.3}$$

where $\langle S' \rangle_r$ denotes the cross-sectional average of entropy oscillations of the inviscid fluid. Also note that W_ν, showing the viscous attenuation per unit time, is absent in w when the inviscid fluid is considered.

A cyclic integral $\oint Y' \, dX'$ represents a signed area when Y' is plotted against X', as shown in Fig. 6.1. When two quantities Y' and X' are oscillating in time with the same frequency, an elliptical orbit is drawn in the $X' - Y'$ plane. The sign of the enclosed area is positive when the rotation direction is clockwise, whereas it is negative when the rotation direction is reversed. Based on this fact, let's consider a three-dimensional elliptical orbit created by the fluid parcel in the $p' - \langle S' \rangle_r - \xi'$ space.

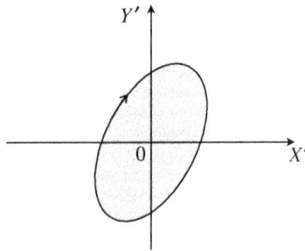

Figure 6.1: $Y'(t) - X'(t)$ diagram when time t changes over a period of oscillations. Enclosed area $\oint Y' \, dX'$ times the frequency gives the time average of the product of Y' and dX'/dt. The sign of the area is positive if the rotation direction is clockwise, and is negative if anticlockwise. The rotation direction reflects the phase relation between Y' and X'. If Y' leads by $90°$, X' reaches the maximum after Y' takes the maximum. If one remembers this simple rule, it would be easy to determine the correct rotation direction.

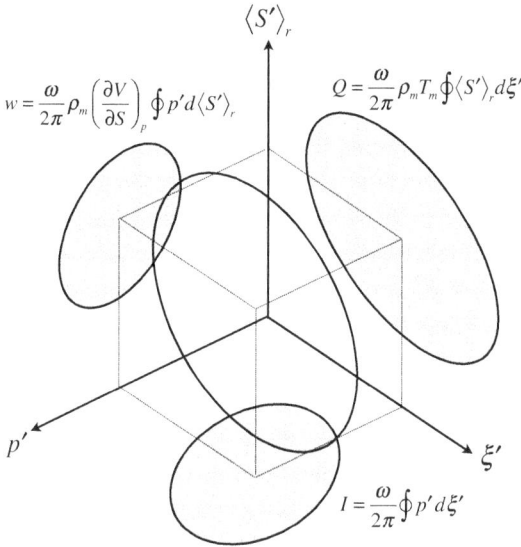

$$\langle S'\rangle_r$$

$$w = \frac{\omega}{2\pi}\rho_m \left(\frac{\partial V}{\partial S}\right)_p \oint p' d\langle S'\rangle_r$$

$$Q = \frac{\omega}{2\pi}\rho_m T_m \oint \langle S'\rangle_r d\xi'$$

$$p'$$

$$\xi'$$

$$I = \frac{\omega}{2\pi}\oint p' d\xi'$$

Figure 6.2: Elliptical orbit created by a fluid parcel in the $p' - \langle S'\rangle_r - \xi'$ space. The signed areas of the projections of the ellipse onto the $p' - \xi'$, $\langle S'\rangle_r - \xi'$, and $\langle S'\rangle_r - p'$ planes corresponds to I, Q, and w, respectively.

Figure 6.2 shows an ellipse created by a fluid parcel oscillating steadily in time. Consider a projection onto the $\langle S'\rangle_r - \xi'$ plane. The signed area times the frequency $[= \omega/(2\pi)]$ yields the entropy flux density s carried by that fluid parcel, from which the heat flux density Q is obtained by taking a product with the mean temperature T_m. The signed area of the projection onto the $p' - \xi'$ plane gives the work flux density I when the frequency is multiplied with it. Also, the work source representing the mutual conversion from Q to I is given by the signed area of the projection onto the $\langle S'\rangle_r - p'$ plane.

The oscillating amplitudes of three quantities p', $\langle S'\rangle_r$, and ξ' are of importance: if the amplitude of $\langle S'\rangle_r$ is zero, the entropy flux density s and the heat flux density Q must be zero, as well as the work source w. Only the work flux density I may be finite. This situation corresponds to the adiabatic sound waves. If p' is zero, then I and w should be zero, so that only Q is possible to exist. On the other hand, if ξ' is zero, $I = Q = 0$, but w may be present. Also the phasing between the three quantities is important. If two of them oscillate in-phase or anti-phase, the projection onto the associated plane becomes a line. As a result, the enclosed area is zero.

The graphical representation using the $p' - \langle S' \rangle_r - \xi'$ space helps to intuitively understand the mechanism of I, Q, and w, as presented by Inoue [64]. His article was accepted by young researchers who had difficulty in understanding Tominaga's original paper [16]. We present the graphical method to understand I, Q, and w, in step by step order, following Inoue's article.

6.2 Visualizing Oscillations of Pressure, Displacement, and Cross-Sectional Mean Entropy

6.2.1 *Pressure oscillation*

Consider a fluid parcel oscillating in a circular tube with positive temperature gradient. The displacement oscillation is taken as a reference when the phasing is considered for other oscillating quantities such as pressure and entropy. When the displacement oscillation $\xi' = \text{Re}[\xi_1 e^{i\omega t}]$ is expressed as

$$\xi' = |\xi_1| \cos(\omega t), \tag{6.4}$$

the pressure oscillation $p' = \text{Re}[p_1 e^{i\omega t}]$ is expressed by

$$p' = |p_1| \cos(\omega t + \theta) \tag{6.5}$$

which can be transformed as

$$p' = |p_1| \cos\theta \cos(\omega t) + |p_1| \sin\theta \cos\left(\omega t + \frac{\pi}{2}\right), \tag{6.6}$$

where θ denotes the phase lead of p' relative to ξ'.[1]

The amplitude of pressure oscillation component that is in phase with the displacement oscillation is $|p_1| \cos\theta$, whereas the component out of phase by $\pi/2$ is $|p_1| \sin\theta$. In a standing wave created in a lossless resonance tube, the phase difference between pressure and velocity is $\pm\pi/2$, as we have seen in Section 3.4. Hence, the phase difference is 0 or π between the pressure and displacement. Therefore,

---

[1]In other words, p_1 is written as $|p_1| e^{i\theta}$.

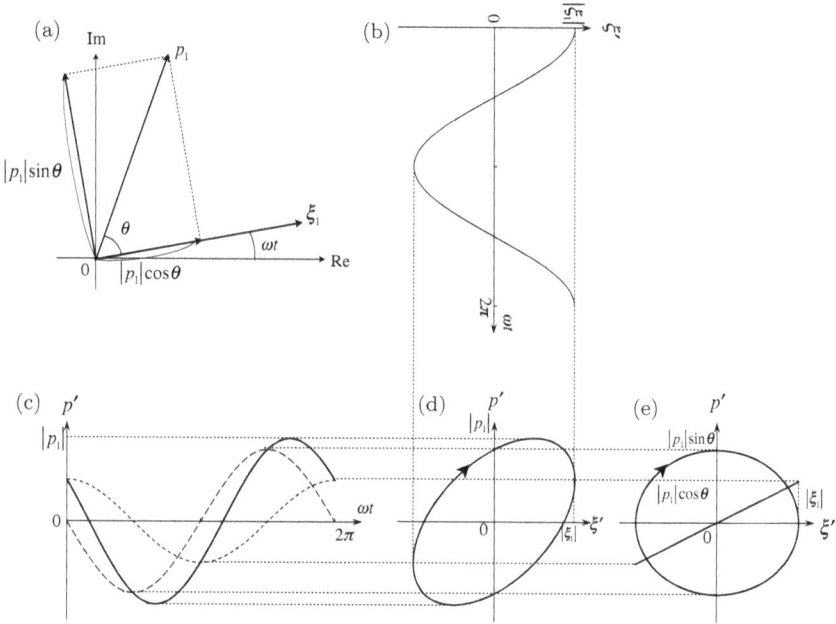

Figure 6.3: Pressure oscillation and displacement oscillation: (a) phasor representation, (b) temporal change of ξ', (c) temporal changes of p', SWC and TWC, (d) orbit in the $p' - \xi'$ plane, and (d) orbits of SWC and TWC in the $p' - \xi'$ plane. SWC and TWC mean the standing wave component and traveling wave component, respectively.

the oscillation component $|p_1| \cos\theta \cos(\omega t)$ is named a *standing wave component* of the pressure oscillation. The phase difference between the pressure and displacement is $\pm\pi/2$ in a traveling wave. Therefore, the oscillation component $|p_1| \sin\theta \cos(\omega t + \pi/2)$ is called a *traveling wave component* (or progressive wave component) of the pressure.

Figure 6.3 (a) illustrates the phasor representation of the complex amplitudes p_1 and ξ_1 when $\theta = \pi/3$. When p_1 is decomposed into a portion in a direction parallel to ξ_1 and normal to ξ_1, each of which gives $|p_1| \cos\theta$ and $|p_1| \sin\theta$, respectively. Figures (b) and (c) show the corresponding temporal changes of ξ' and p', which is combined to create a $p' - \xi'$ diagram in (d). The orbits drawn by the standing wave component and traveling wave component of p' and ξ' is depicted in (e), which are an ellipse and a line, respectively.

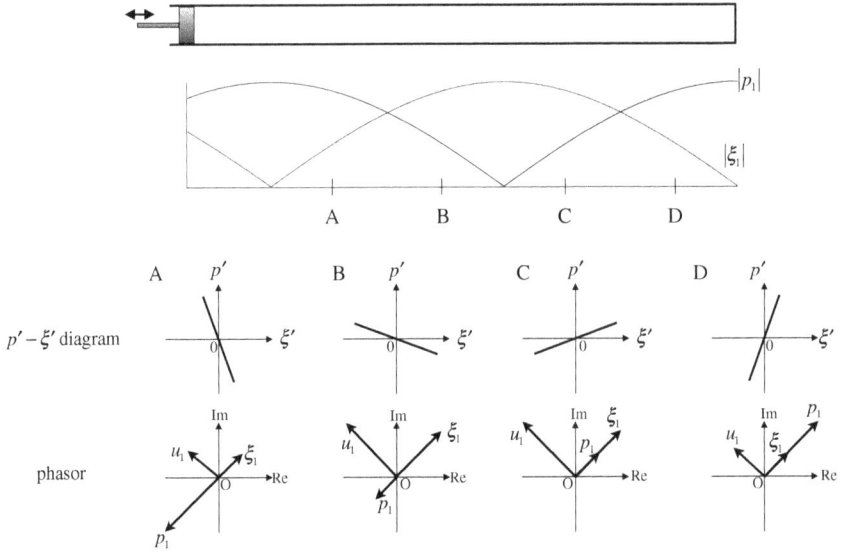

Figure 6.4: $p' - \xi'$ diagrams of a standing wave field in a resonance tube.

The $p' - \xi'$ diagram is specific to each of the fluid parcels under consideration. Namely, the diagram reflects the local acoustic field. Recall the acoustic field of the resonance tube closed by a plate. The corresponding $p' - \xi'$ diagrams are shown in Fig. 6.4. The slope of the line orbit is steep in positions A and D near the pressure antinode, whereas it is gentle in B and C near the pressure node. Also the sign of the slope changes as to whether the location of the pressure maximum is on the right hand side or the left. Figure 6.5 shows the $p' - \xi'$ diagrams in a traveling wave propagating in a positive direction of x. In this case, the orbits are the same ellipse rotating in a clockwise direction independently of the position. In a backward traveling wave, the orbit is also an ellipse, but the rotation direction is reversed.

6.2.2 *Cross-sectional mean entropy oscillation*

Consider the cross-sectional mean entropy oscillation

$$\langle S' \rangle_r = \mathrm{Re}[\langle S_1 \rangle_r e^{i\omega t}].$$

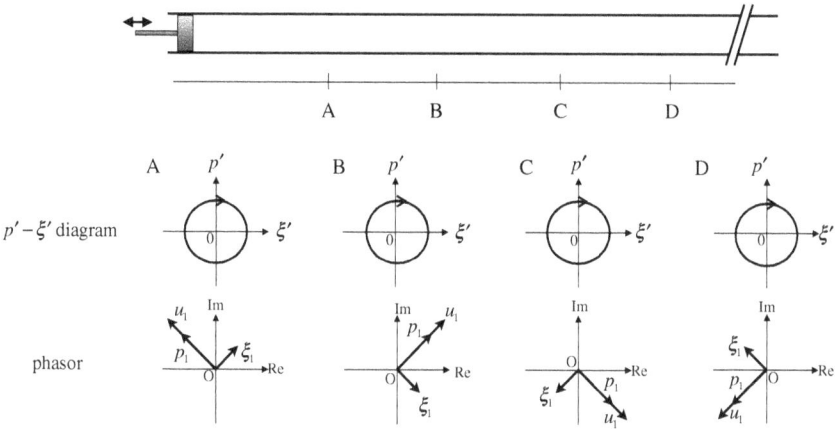

Figure 6.5: $p' - \xi'$ diagrams of a traveling wave field in a reflection-less tube.

The complex amplitude $\langle S_1 \rangle_r$ is expressed as

$$\langle S_1 \rangle_r = \chi_\alpha S_{1L}$$

for an inviscid fluid. Here, S_{1L} represents the entropy oscillation of a gas parcel moving back and fourth along the tube wall while keeping good thermal contact with it. It is written as

$$S_{1L} = \left(\frac{\partial S}{\partial p}\right)_T p_1 + \left(\frac{\partial S}{\partial T}\right)_p \frac{dT_m}{dx}\xi_1$$

where the first term on the right-hand side represents the entropy oscillation due to isothermal pressure oscillation, whereas the second term means the entropy oscillation caused by displacement oscillation along the tube having a non-zero temperature gradient. The cross-sectional mean entropy oscillation $\langle S' \rangle_r$ is given by

$$\langle S' \rangle_r = \mathrm{Re}\left[\left(\frac{\partial S}{\partial p}\right)_T \chi_\alpha p_1 e^{i\omega t}\right] + \mathrm{Re}\left[\left(\frac{\partial S}{\partial T}\right)_p \frac{dT_m}{dx}\chi_\alpha \xi_1 e^{i\omega t}\right].$$

$$(6.7)$$

Hence, $\langle S' \rangle_r$ is decomposed into a portion proportional to $\chi_\alpha p_1$ and that to $\chi_\alpha \xi_1$.

By explicitly writing the complex quantity χ_α as $\chi_\alpha = \chi_\alpha' + i\chi_\alpha''$, the entropy oscillation due to pressure oscillation is expressed by

$$
\mathrm{Re}\left[\left(\frac{\partial S}{\partial p}\right)_T \chi_\alpha p_1 e^{i\omega t}\right]
$$

$$
= \left(\frac{\partial S}{\partial p}\right)_T |p_1|\, \mathrm{Re}\left[(\chi_\alpha' + i\chi_\alpha'')e^{i(\omega t+\theta)}\right]
$$

$$
= \left(\frac{\partial S}{\partial p}\right)_T |p_1|\left[\chi_\alpha' \cos(\omega t + \theta) + \chi_\alpha'' \cos\left(\omega t + \frac{\pi}{2} + \theta\right)\right]. \quad (6.8)
$$

Therefore, it is given by a sum of oscillations in phase with p' and out of phase by $\pi/2$ with p'. It should be noted that χ_α'' is negative; the second term represents the oscillation delayed by $\pi/2$ relative to p'.

Figure 6.6 (a) represents a phasor plot of p_1 and components of pressure-induced entropy oscillation $(\partial S/\partial p)_T \chi_\alpha p_1$, where we have

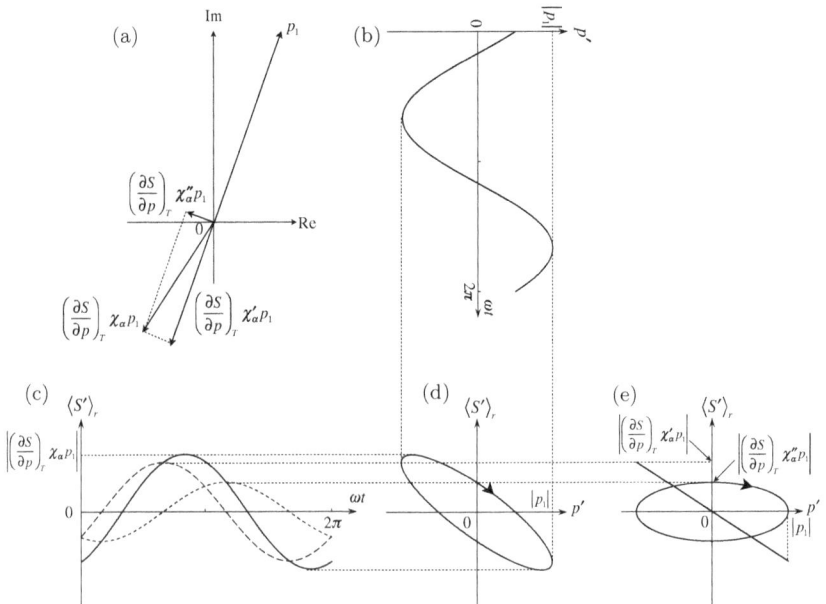

Figure 6.6: Pressure oscillation and entropy oscillation due to pressure: (a) phasor representation, (b) temporal change of p', (c) temporal changes of $\langle S'\rangle_r$, (d) orbit in the $\langle S'\rangle_r - p'$ plane, and (e) orbits of components of entropy oscillation in phase and out of phase with p' in the $\langle S'\rangle_r - p'$ plane.

used that $(\partial S / \partial p)_T$ is negative for normal fluid. Figures (b) and (c) represent the temporal changes of pressure and pressure-induced entropy oscillations. The resulting $\langle S' \rangle_r - p'$ diagrams are depicted in (d) and (e).

In the same way, the entropy oscillation induced by displacement along the temperature gradient can be decomposed as

$$\text{Re} \left[\left(\frac{\partial S}{\partial T} \right)_p \frac{dT_m}{dx} \chi_\alpha \xi_1 e^{i\omega t} \right]$$

$$= \left(\frac{\partial S}{\partial T} \right)_p \frac{dT_m}{dx} |\xi_1| \left[\chi'_\alpha \cos \omega t + \chi''_\alpha \cos \left(\omega t + \frac{\pi}{2} \right) \right]. \tag{6.9}$$

Because $\chi''_\alpha < 0$, the second term on the right-hand side of Eq. (6.9) represents the oscillation delayed by $\pi / 2$ relative to ξ'.

Figure 6.7 illustrates a phasor plot of ξ_1 and components of entropy oscillations induced by displacement along the temperature

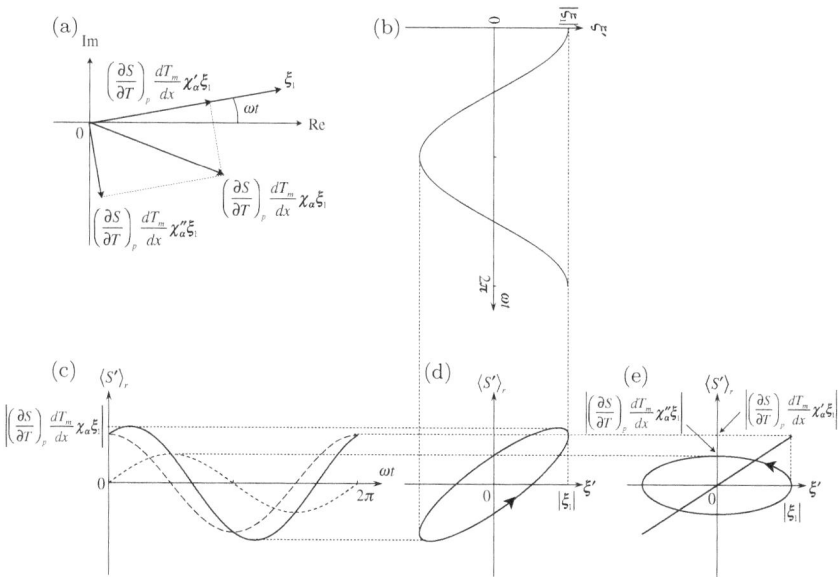

Figure 6.7: Displacement oscillation and entropy oscillation due to displacement: (a) phasor representation, (b) temporal change of ξ', (c) temporal changes of $\langle S' \rangle_r$, (d) orbit in the $\langle S' \rangle_r - \xi'$ plane, and (e) orbits of components of entropy oscillation in phase and out of phase with ξ' in the $\langle S' \rangle_r - \xi'$ plane.

gradient. Figures (b) and (c) represent the temporal changes of displacement and displacement-induced entropy oscillations. The resulting $\langle S' \rangle_r - \xi'$ diagrams are depicted in (d) and (e).

6.3 Work Flux Density

The work flux density I for an inviscid fluid is expressed by

$$ I = \frac{\omega}{2\pi} \oint p' d\xi'. $$

Therefore, I is equal to the area enclosed by a gas parcel on a $p' - \xi'$ diagram times the frequency $f = \omega/(2\pi)$. Figure 6.8 illustrates the $p' - \xi'$ diagram, as well as those drawn by decomposing p' into the traveling wave component (A1) and the standing wave component (A2).

The ellipse depicted in Fig. 6.8 A1 represents the work flux density going in a positive direction of x, as the trace of the gas parcel goes clockwise on the ellipse. The signed area of the ellipse is given by $\pi |p_1||\xi_1| \sin\theta$. On the other hand, the trace becomes a line segment in A2, where the standing wave component of pressure is drawn against ξ', and the enclosed area results in zero. Therefore, only the traveling

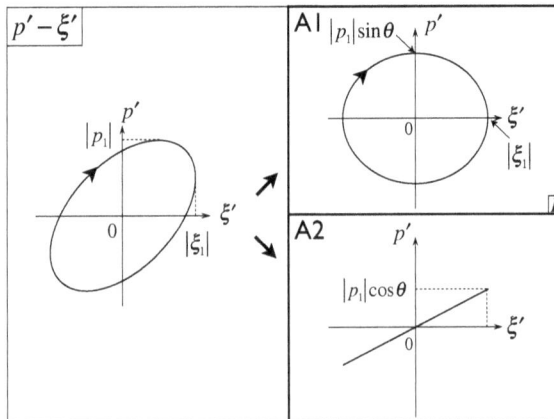

Figure 6.8: Diagrams of pressure oscillation p' and displacement oscillation ξ'.

wave component contributes to the work flux density:

$$I = \frac{\omega}{2}|p_1||\xi_1|\sin\theta. \tag{6.10}$$

If we use the complex velocity amplitude $u_1 = i\omega\xi_1$, I is written as

$$I = \frac{1}{2}|p_1||u_1|\cos\phi, \tag{6.11}$$

where ϕ ($= \pi/2 - \theta$) means the phase of u_1 relative to p_1 (See Fig. 6.13).

6.4 Heat Flux Density

The heat flux density Q for an inviscid fluid is expressed by

$$Q = \frac{\omega}{2\pi}\rho_m T_m \oint \langle S'\rangle_r d\xi',$$

which is equal to the area enclosed by a gas parcel on a $\langle S'\rangle_r - \xi'$ diagram times a factor of $\omega\rho_m T_m/(2\pi)$. As shown in Eq. (6.7), $\langle S'\rangle_r$ consists of a component due to pressure oscillation and that due to displacement oscillation. Owing to this fact, classification of Q is possible depending on the components of entropy oscillation. We first investigate the heat flux densities Q_{prog} and Q_{stand} caused by entropy oscillation due to pressure oscillation, and then consider Q_D that is produced by entropy oscillation due to displacement oscillation.

6.4.1 *Heat flux density component due to pressure oscillation* (Q_{prog} *and* Q_{stand})

The left panel of Fig. 6.9 shows $\langle S'\rangle_r - p'$ diagrams of Figs. 6.6 (d) and (e). The top panel shows $p' - \xi'$ diagrams in Fig. 6.8. By combining these panels, we can create $\langle S'\rangle_r - \xi'$ diagrams B1, B2, C1, and C2 as shown in the bottom right panel.

Diagram B1 presents pressure-induced entropy oscillation that is in-phase with the traveling wave component of p' and the displacement oscillation ξ'. The traveling wave component of p' rapidly decreases when the gas parcel is positioned near the left end. Because the gas parcel experiences isothermal pressure change, the

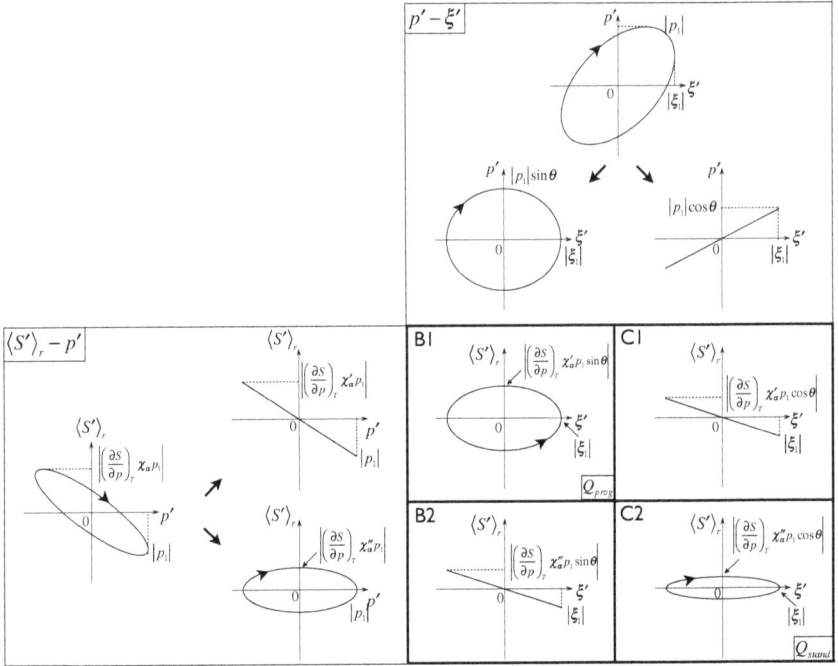

Figure 6.9: Diagram showing p'-induced component of entropy oscillation $\langle S'\rangle_r$ and p' (left) and diagram of $p' - \xi'$ (top). The diagrams B1, B2, C1, and C2 are obtained by combining the traveling-wave and standing-wave components of p' in the top panel with in-phase and out-of-phase components of $\langle S'\rangle_r$ in the left panel.

increase of the entropy reflects the heat transfer from the channel wall to the gas parcel. The gas parcels then move to the left and reduce the entropy through the isothermal pressure increase by heat transfer to the wall. Since the heat transfers from and to the channel wall occur in different positions, the axial heat transport takes place from the right to the left in the same way as a bucket brigade. The signed area enclosed by the gas parcel is equal to

$$-\pi\left\{-\left(\frac{\partial S}{\partial p}\right)_T \chi'_\alpha |p_1| \sin\theta\right\}|\xi_1| = \pi\left(\frac{\partial S}{\partial p}\right)_T \chi'_\alpha |p_1||\xi_1| \sin\theta.$$

By taking a product with $\omega\rho_m T_m/(2\pi)$, one obtains the heat flux density component Q_{prog} due to the traveling wave component of the pressure:

$$Q_{prog} = \frac{\omega}{2} \rho_m T_m \left(\frac{\partial S}{\partial p} \right)_T \chi'_\alpha |p_1||\xi_1| \sin \theta. \qquad (6.12)$$

The thermal expansion coefficient, $\beta = \rho_m (\partial V / \partial T)_p$ satisfies a thermodynamic relation

$$\rho_m \left(\frac{\partial S}{\partial p} \right)_T = -\beta. \qquad (6.13)$$

Therefore, Q_{prog} is further transformed as

$$Q_{prog} = -\frac{\omega}{2} \chi'_\alpha \beta T_m |p_1||\xi_1| \sin \theta. \qquad (6.14)$$

We see a simple relation between Q_{prog} and I

$$Q_{prog} = -\chi'_\alpha \beta T_m I, \qquad (6.15)$$

which means that Q_{prog} flows in an opposite direction to I with its magnitude proportional to I. Also, it should be noticed that Q_{prog} originates from isothermal heat transfer with the channel wall, as it is proportional to χ'_α. The subscript of Q_{prog} indicates that it is caused by the traveling wave component of pressure oscillation.

Diagram B2 presents pressure-induced entropy oscillation that is out-of-phase by $\pi/2$ with the traveling wave component of p' and the displacement oscillation ξ'. As a result of the phase difference of $\pi/2$, this component of entropy oscillation is in phase with displacement oscillation. Therefore, the heat absorption and release by the gas parcel take place at the same position, and hence, the entropy-displacement diagram becomes a line segment, leading to zero contribution to the heat flux density Q.

Diagram C1 presents the pressure-induced entropy oscillation that is in phase with standing wave component of p' and the displacement oscillation ξ'. The entropy of the gas parcel changes in phase with the standing wave component of pressure, and hence it oscillates in phase with ξ'. Therefore, the diagram becomes a line segment, which means that this component of entropy oscillation does not contribute to the heat flux density Q.

Diagram C2 presents the pressure-induced entropy oscillation that is out-of-phase by $\pi/2$ with standing wave component of p' and displacement oscillation ξ'. This component of entropy oscillation leads the pressure oscillation by $\pi/2$, and hence, the entropy rapidly decreases when the gas parcel reaches the right end, where pressure is at the maximum. In the same way, the entropy increases when the gas parcel reaches the left end. Such entropy oscillation means that the heat absorption and release takes place in different positions. Therefore, the gas parcel maintains axial heat transport by the bucket brigade of the entropy. The signed area enclosed by the gas parcel is given by

$$\pi \left(\frac{\partial S}{\partial p}\right)_T \chi_\alpha'' |p_1||\xi_1| \cos\theta.$$

By multiplying a factor of $\omega\rho_m T_m/(2\pi)$, one obtains the heat flux density Q_{stand}:

$$Q_{stand} = \frac{\omega}{2}\rho_m T_m \left(\frac{\partial S}{\partial p}\right)_T \chi_\alpha'' |p_1||\xi_1| \cos\theta. \tag{6.16}$$

Use of Eq. (6.13) gives

$$Q_{stand} = -\frac{\omega}{2}\chi_\alpha'' \beta T_m |p_1||\xi_1| \cos\theta. \tag{6.17}$$

The heat flux density Q_{stand} flows in the direction where the pressure becomes large. This direction has nothing to do with the sign of the temperature gradient nor the sign of the work flux density I. The heat flux density Q_{stand} is proportional to χ_α'', which means that it is caused by irreversible heat exchange processes with the channel walls. The subscript *stand* means that Q_{stand} originates from the standing wave component of p'.

6.4.2 *Heat flux density component due to displacement oscillation* (Q_D)

The left panel of Fig. 6.10 shows the $\langle S'\rangle_r - \xi'$ diagram presented in Fig. 6.7 (d), where entropy oscillation caused by the displacement along the temperature gradient is given. Diagram D1 on the top right

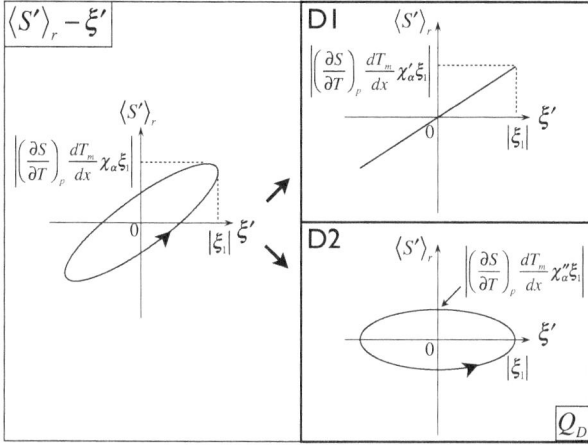

Figure 6.10: Diagram showing ξ'-induced component of entropy oscillation $\langle S' \rangle_r$ and ξ' (left). Diagrams D1 and D2 show the $\langle S' \rangle_r - \xi'$ diagrams of entropy oscillation components in phase with ξ' and out-of-phase by $\pi/2$ with ξ'.

of Fig. 6.10 presents the displacement-induced entropy oscillation that is in phase with the displacement oscillation ξ'. Although the entropy of the gas parcel changes with time owing to the heat exchange with the channel wall, the bucket brigade of entropy does not take place. Therefore, there is no contribution to the heat flux density.

Diagram D2 on the bottom right of Fig. 6.10 presents the displacement-induced entropy oscillation that is out-of-phase by $\pi/2$ with ξ'. Because of the phase delay, the entropy keeps going up during the time interval when $\xi' > 0$; it decreases while $\xi' < 0$. As a result, axial heat transport is achieved from the right-hand side (hot-end side) to the left-hand side (cold-end side) by a bucket brigade of entropy. The signed area enclosed by the gas parcel is given by

$$-\pi \left\{ -\left(\frac{\partial S}{\partial T} \right)_p \frac{dT_m}{dx} \chi_\alpha'' |\xi_1| \right\} |\xi_1| = \pi \left(\frac{\partial S}{\partial T} \right)_p \frac{dT_m}{dx} \chi_\alpha'' |\xi_1|^2.$$

By multiplying a factor of $\omega \rho_m T_m/(2\pi)$ yields the heat flux density Q_D due to the displacement oscillation along the temperature gradient:

$$Q_D = \frac{\omega}{2}\chi_\alpha'' \rho_m T_m \left(\frac{\partial S}{\partial T}\right)_p \frac{dT_m}{dx}|\xi_1|^2. \tag{6.18}$$

Because the isobaric specific heat C_p is given by

$$C_p = T_m \left(\frac{\partial S}{\partial T}\right)_p, \tag{6.19}$$

we have a simpler expression for Q_D:

$$Q_D = \frac{\omega}{2}\chi_\alpha'' \rho_m C_p \frac{dT_m}{dx}|\xi_1|^2. \tag{6.20}$$

Since Q_D is proportional to χ_α'', it arises from the irreversible heat exchange processes with the wall. Also, it should be noted that Q_D always flows down the temperature gradient because $\chi_\alpha'' < 0$. In other words, Q_D introduces heat loss in a heat pump, while it serves as the main heat flux component in a dream pipe. The subscript D represents displacement and/or dream pipe.

6.4.3 *Summary of heat flux density components*

For an inviscid fluid, the heat flux density Q maintained by oscillating fluid parcels is decomposed into

$$Q = Q_{prog} + Q_{stand} + Q_D, \tag{6.21}$$

where

$$Q_{prog} = -\frac{\omega}{2}\chi_\alpha' \beta T_m |p_1||\xi_1| \sin\theta,$$

$$Q_{stand} = -\frac{\omega}{2}\chi_\alpha'' \beta T_m |p_1||\xi_1| \cos\theta,$$

$$Q_D = \frac{\omega}{2}\chi_\alpha'' \rho_m C_p \frac{dT_m}{dx}|\xi_1|^2.$$

If we use the complex velocity amplitude $u_1 = i\omega\xi_1$, in place of ξ_1, the heat flux density components are rewritten as follows:

$$Q_{prog} = -\frac{1}{2}\chi'_\alpha \beta T_m |p_1||u_1|\cos\phi, \tag{6.22}$$

$$Q_{stand} = -\frac{1}{2}\chi''_\alpha \beta T_m |p_1||u_1|\sin\phi, \tag{6.23}$$

$$Q_D = \frac{1}{2\omega}\chi''_\alpha \rho_m C_p \frac{dT_m}{dx}|u_1|^2, \tag{6.24}$$

where ϕ denotes the phase lead of u_1 relative to p_1 ($\phi = \pi/2 - \theta$).

Let us summarize the properties of Q_{prog}, Q_{stand}, and Q_D.

(1) As shown in Table 6.1, Q_{prog} is caused by the traveling wave component of pressure, $|p_1|\sin\theta$, through the reversible heat exchange processes of the gas with the wall as it is proportional to χ'_α, whereas Q_{stand} originates from the standing wave component of pressure, $|p_1|\cos\theta$, through the irreversible processes represented by χ''_α. As we will discuss in Chapter 8, Q_{prog} plays an essential role in a looped tube acoustic cooler, GM refrigerator, and pulse tube cooler; Q_{stand} becomes important in a resonance tube cooler. The heat flux density component Q_D is caused by irreversible heat exchange processes, as it is proportional to χ''_α.

(2) Both Q_{prog} and Q_{stand} are proportional to a product of thermal expansion coefficient β and mean temperature T_m. In an ideal gas, a relation $\beta T_m = 1$ holds. Usually, β of a gas is much larger than that of a liquid. Therefore, Q_{prog} and Q_{stand} are more enhanced in gases than in liquids. On the other hand, Q_D is proportional to a product of mean density ρ_m and isobaric specific heat C_p, it becomes much larger in liquids than in gases.

Table 6.1: Heat flux density components.

		Thermodynamic process	
		χ'_α	χ''_α
	$|p_1||\xi_1|\sin\theta$	Q_{prog}	—
Amplitude factor	$|p_1||\xi_1|\cos\theta$	—	Q_{stand}
	$|\xi_1|^2$	—	Q_D

Therefore, the dream pipe usually employs liquids as the working fluid.

(3) Since Q_D is proportional to $|u_1|^2/\omega = \omega|\xi_1|^2$, it increases quadratically with the displacement amplitude. However, Q_D has no relation with the pressure amplitude $|p_1|$ nor the phase relation between pressure and displacement.

6.5 Work Source

The work source w for an inviscid fluid is expressed by

$$w = \frac{\omega}{2\pi}\rho_m \left(\frac{\partial V}{\partial S}\right)_p \oint p'd\langle S'\rangle_r.$$

Here the integral $(\partial V/\partial S)_p \oint p'd\langle S'\rangle_r$ is equal to $\oint p'd\langle V'\rangle_r$, because we have

$$\langle V'\rangle_r = \left(\frac{\partial V}{\partial p}\right)_S p' + \left(\frac{\partial V}{\partial S}\right)_p \langle S'\rangle_r, \qquad (6.25)$$

and $\oint p'dp' = 0$. Therefore, w means the energy conversion rate done by the gas parcel per unit volume and time. The coefficient $(\partial V/\partial S)_p$ is always positive, since

$$\left(\frac{\partial V}{\partial S}\right)_p = \frac{K_T - K_S}{\beta} > 0, \qquad (6.26)$$

where K_T is the isothermal compressibility and K_S the adiabatic compressibility. Thus, the gas parcel generates work from heat when $p'd\langle S'\rangle_r > 0$, whereas it absorbs work when $p'd\langle S'\rangle_r < 0$.

As shown in Eq. (6.7), the cross-sectional mean entropy oscillation, $\langle S'\rangle_r$, has two components; one is caused by pressure oscillation, the other is due to displacement oscillation:

$$\langle S'\rangle_r = \mathrm{Re}\left[\left(\frac{\partial S}{\partial p}\right)_T \chi_\alpha p_1 e^{i\omega t}\right] + \mathrm{Re}\left[\left(\frac{\partial S}{\partial T}\right)_p \frac{dT_m}{dx}\chi_\alpha \xi_1 e^{i\omega t}\right].$$

Accordingly, the work source w can be decomposed into

$$w = W_p + W_{prog} + W_{stand}$$

where W_p is associated with the entropy oscillation caused by pressure oscillation, whereas W_{prog} and W_{stand} originate from the entropy oscillation caused by displacement oscillation.

6.5.1 *Work source component due to pressure oscillation (W_p)*

Figure 6.11 illustrates the relation between pressure oscillation p' and entropy oscillation $\langle S' \rangle_r$ due to p'. Diagram E1 depicts specifically the relation between the pressure oscillation and entropy oscillation that is in phase with p' without phase delay. As stated above, $p' - \langle S' \rangle_r$ diagram equivalently serves as $p' - \langle V' \rangle_r$. Therefore the line trace in E1 means that the gas parcel generates no net work.

The diagram E2 depicts the relation between the pressure oscillation and entropy oscillation that is out of phase with p' by $\pi/2$. The gas parcel traces a circular orbit in anticlockwise direction. The signed area in E2 is given by

$$-\pi|p_1|\left(\frac{\partial S}{\partial p}\right)_T \chi_\alpha''|p_1| = -\pi\left(\frac{\partial S}{\partial p}\right)_T \chi_\alpha''|p_1|^2.$$

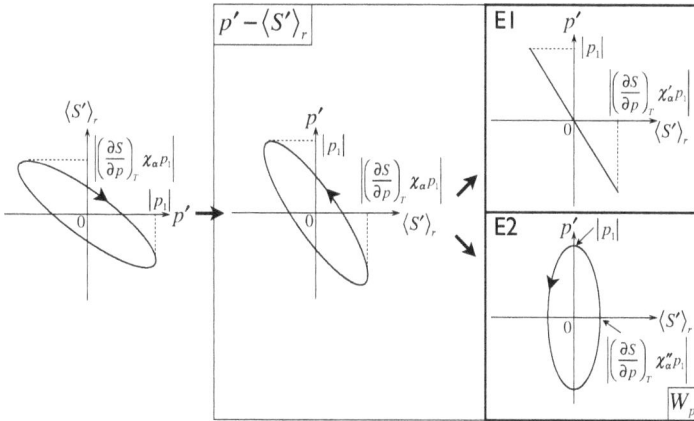

Figure 6.11: Pressure oscillation p' and entropy oscillation $\langle S' \rangle_r$ due to p'. Diagrams E1 and E2 show the entropy oscillations in phase with p' and out-of-phase by $\pi/2$ with p'.

By multiplying $\omega/(2\pi)\rho_m(\partial V/\partial S)_p$, we have the work source component W_p:

$$W_p = -\frac{\omega}{2}\rho_m \left(\frac{\partial V}{\partial S}\right)_p \left(\frac{\partial S}{\partial p}\right)_T \chi_\alpha''|p_1|^2.$$

From the thermodynamic relations in Eqs. (6.13) and (6.26), W_p is also expressed as

$$W_p = \frac{\omega}{2}(K_T - K_S)\chi_\alpha''|p_1|^2. \tag{6.27}$$

Since W_p is proportional to χ_α'', it is caused by irreversible heat exchange between the gas and the tube wall. Indeed, as χ_α'' is always negative, W_p is also negative, showing energy dissipation. The subscript p reminds that W_p is proportional to $|p_1|^2$.

6.5.2 *Work source components due to displacement oscillation (W_{prog} and W_{stand})*

The diagrams of $p' - \langle S'\rangle_r$ for work source components W_{prog} and W_{stand} can be drawn by combining the diagrams of $p' - \xi'$ and $\langle S'\rangle_r - \xi'$, as shown in Fig. 6.12. Diagram F1 shows the traveling wave component of p' and entropy oscillation $\langle S'\rangle_r$ that is in phase with ξ'. The signed area of the trace in F1 is given by

$$\pi \left(\frac{\partial S}{\partial T}\right)_p \frac{dT_m}{dx} \chi_\alpha'|\xi_1||p_1|\sin\theta.$$

Multiplying $\omega/(2\pi)\rho_m(\partial V/\partial S)_p$ gives work source component

$$W_{prog} = \frac{\omega}{2}\rho_m \left(\frac{\partial V}{\partial S}\right)_p \left(\frac{\partial S}{\partial T}\right)_p \frac{dT_m}{dx}\chi_\alpha'|p_1||\xi_1|\sin\theta. \tag{6.28}$$

From the formula of partial derivatives, we have

$$\left(\frac{\partial V}{\partial S}\right)_p \left(\frac{\partial S}{\partial T}\right)_p = \left(\frac{\partial V}{\partial T}\right)_p, \tag{6.29}$$

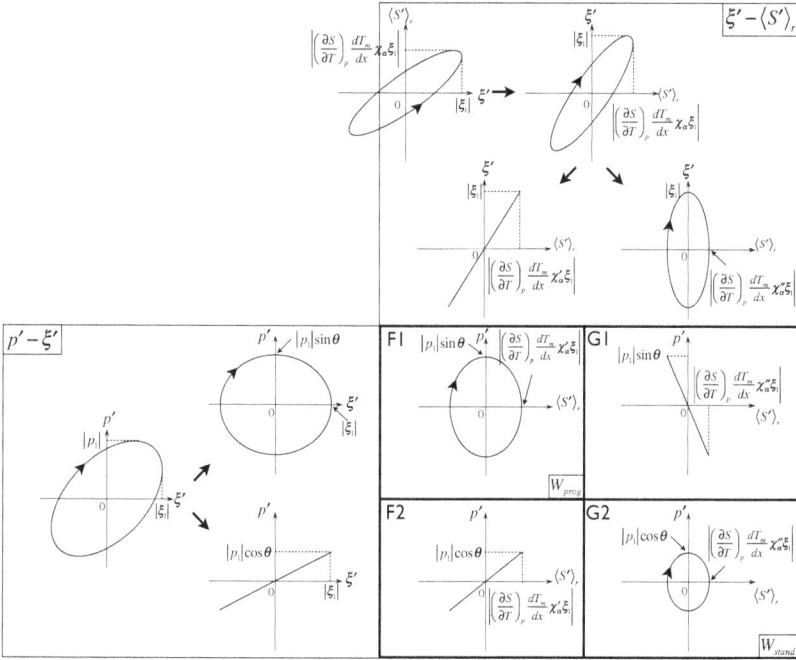

Figure 6.12: Pressure oscillation p' and entropy oscillation $\langle S' \rangle_r$ due to displacement oscillation ξ'. Diagram F1 (F2) are drawn on the basis of the traveling wave component (the standing wave component) of p' and entropy oscillation $\langle S' \rangle_r$ that is in phase with ξ'. Diagram G1 (G2) are drawn on the basis of the traveling wave component (the standing wave component) of p' and entropy oscillation $\langle S' \rangle_r$ that is out-of-phase by $\pi/2$ with ξ'.

which is further rewritten by using the thermal expansion coefficient β as

$$\left(\frac{\partial V}{\partial T}\right)_p = \frac{\beta}{\rho_m}.$$

Therefore, W_{prog} is transformed as

$$W_{prog} = \frac{\omega}{2}\beta\frac{dT_m}{dx}\chi'_\alpha |p_1||\xi_1|\sin\theta. \qquad (6.30)$$

It is easy to see that W_{prog} is related to the work flux density I by

$$W_{prog} = \beta\frac{dT_m}{dx}\chi'_\alpha I. \qquad (6.31)$$

Work source W_{prog} originates from the reversible heat exchange process between the gas and the tube wall as it is in proportion to χ'_α. Since χ'_α is positive, W_{prog} is positive when the signs of I and dT_m/dx are equal to each other, namely when I goes up the temperature gradient. Therefore, the regenerator serves as the amplifier of I when W_{prog} is strong. This is why the engine is called a traveling wave engine when I is amplified by the temperature gradient.

Diagram F2 is made of the standing wave component of p' and entropy oscillation $\langle S'\rangle_r$ that is in phase with ξ'. The trace of the gas parcel is a line. Therefore, no energy conversion takes place, although both pressure oscillation p' and entropy oscillation $\langle S'\rangle_r$ are not zero.

Diagram G1 is drawn by the traveling wave component of p' and entropy oscillation $\langle S'\rangle_r$ that is out of phase with ξ' by $\pi/2$. The trace of the gas parcel is a line. Therefore, no energy conversion takes place.

Diagram G2 is drawn by the standing wave component of p' and entropy oscillation $\langle S'\rangle_r$ that is out of phase with ξ' by $\pi/2$. The signed area of the trace in G2 is given by

$$\pi\left[-\left(\frac{\partial S}{\partial T}\right)_p \frac{dT_m}{dx}\chi''_\alpha|\xi_1|\right]|p_1|\cos\theta = -\pi\left(\frac{\partial S}{\partial T}\right)_p \frac{dT_m}{dx}\chi''_\alpha|p_1||\xi_1|\cos\theta.$$

Multiplying $\omega/(2\pi)\rho_m(\partial V/\partial S)_p$ gives work source component

$$W_{stand} = -\frac{\omega}{2}\rho_m\left(\frac{\partial V}{\partial S}\right)_p\left(\frac{\partial S}{\partial T}\right)_p \frac{dT_m}{dx}\chi''_\alpha|p_1||\xi_1|\cos\theta. \qquad (6.32)$$

It is rewritten by using Eq. (6.29) and β as

$$W_{stand} = -\frac{\omega}{2}\beta\frac{dT_m}{dx}\chi''_\alpha|p_1||\xi_1|\cos\theta, \qquad (6.33)$$

which is also transformed as

$$W_{stand} = \frac{Q_{stand}}{T_m}\frac{dT_m}{dx}. \qquad (6.34)$$

We see that sign of W_{stand} is equal to that of Q_{stand} when the temperature gradient is positive. Therefore, in the case when the pressure goes up as the gas parcel moves to the hot part, W_{stand} becomes positive, meaning that the energy conversion tends to increase I. However, W_{stand} has nothing to do with I. For example, when $\theta = 0$, $I = 0$ but W_{stand} can be positive when dT_m/dx is positive. As a result, I changes its sign from negative to positive at ends of the stack when W_{stand} dominates the energy conversion. This is why the engine is called a standing wave engine when I is emitted from both sides of the stack.

6.5.3 *Summary of work source components*

For an inviscid fluid, the work source w maintained by oscillating fluid parcels is decomposed into

$$w = W_p + W_{prog} + W_{stand}, \qquad (6.35)$$

where

$$W_p = \frac{\omega}{2}(K_T - K_S)\chi_\alpha''|p_1|^2,$$

$$W_{prog} = \frac{\omega}{2}\beta\frac{dT_m}{dx}\chi_\alpha'|p_1||\xi_1|\sin\theta,$$

$$W_{stand} = -\frac{\omega}{2}\beta\frac{dT_m}{dx}\chi_\alpha''|p_1||\xi_1|\cos\theta.$$

If we use the complex velocity amplitude $u_1 = i\omega\xi_1$, in place of ξ_1, work source components W_{prog} and W_{stand} are rewritten as follows:

$$W_{prog} = \frac{1}{2}\beta\frac{dT_m}{dx}\chi_\alpha'|p_1||u_1|\cos\phi, \qquad (6.36)$$

$$W_{stand} = -\frac{1}{2}\beta\frac{dT_m}{dx}\chi_\alpha''|p_1||u_1|\sin\phi, \qquad (6.37)$$

where ϕ denotes the phase lead of u_1 relative to p_1 ($\phi = \pi/2 - \theta$).

Let us summarize the properties of W_{prog}, W_{stand}, and W_p.

(1) As shown in Table 6.2, W_{prog} is caused by the traveling wave component of pressure, $|p_1|\sin\theta$, through the reversible heat exchange processes of the gas with the wall as it is proportional to χ_α'; W_{stand} originates from the standing wave component of pressure, $|p_1|\cos\theta$, through the irreversible processes represented by χ_α''. As we will discuss in Chapter 7, W_{prog} plays an essential

Table 6.2: Work source components.

		Thermodynamic process	
		χ'_α	χ''_α
Amplitude factor	$|p_1||\xi_1|\sin\theta$	W_{prog}	—
	$|p_1||\xi_1|\cos\theta$	—	W_{stand}
	$|p_1|^2$	—	W_p

role in a looped tube acoustic engine and Stirling engines; W_{stand} becomes important in a resonance tube engine. Work source component W_p is proportional to $|p_1|^2$ and χ''_α, meaning that W_p is associated with the irreversible heat exchange processes and always causes damping of I.

(2) Both W_{prog} and W_{stand} are proportional to a product of thermal expansion coefficient β. In an ideal gas, a relation $\beta T_m = 1$ holds. Usually, β of a gas is much larger than that of a liquid. Therefore, W_{prog} and W_{stand} are small in liquids.

(3) Both W_{prog} and W_{stand} are proportional to the temperature gradient dT_m/dx. Hence, they should vanish in a uniform temperature flow channel. On the other hand, W_p has no relation to the temperature gradient.

6.6 Energy Flow Density and Work Source for Viscous Fluid

6.6.1 *Mathematical formula for a product of oscillating quantities*

Thermoacoustic theory discusses viscous fluid as well as inviscid fluid for a more general description of thermoacoustic phenomena. The physical concept of energy flux density and work source are basically the same as those for the inviscid fluid, but mathematical treatment for viscous fluid becomes a bit complicated, particularly when the time average of two oscillating quantities are involved. Therefore, in this section, we show some mathematical formulas for later use.

Consider here two oscillating quantities $X' = |X_1|\cos(\omega t + \Phi)$ and $Y' = |Y_1|\cos\omega t$. They are expressed as $X' = \text{Re}[X_1 e^{i\omega t}]$ and

$Y' = \mathrm{Re}[Y_1 e^{iwt}]$, respectively. As the phase of X_1 leads that of Y_1 by Φ, a product of X_1 and the complex conjugate $Y_1^\dagger$ of Y_1 is written as

$$X_1 Y_1^\dagger = |X_1||Y_1| \cos \Phi + i|X_1||Y_1| \sin \Phi.$$

The temporal mean of the product $X'Y'$ is given by

$$\langle X'Y' \rangle_t = \frac{1}{2}|X_1||Y_1| \cos \Phi.$$

Therefore, we have a mathematical formula

$$\langle X'Y' \rangle_t = \frac{1}{2}\mathrm{Re}[X_1 Y_1^\dagger]. \tag{6.38}$$

It should be noted that Eq. (6.38) is satisfied for arbitral initial phase of Y'.

Another mathematical formula that is often used in the later part of Section 6.6 is concerned with the temporal mean of X' and dY'/dt. The formula is written as

$$\left\langle X' \frac{dY'}{dt} \right\rangle_t = \frac{1}{2}\omega\mathrm{Im}[X_1 Y_1^\dagger]. \tag{6.39}$$

The derivation is as follows. If we express $X_1 Y_1^\dagger$ by $Z = Z_\mathrm{R} + iZ_\mathrm{I}$, we have

$$\left\langle X' \frac{dY'}{dt} \right\rangle_t = \frac{1}{2}\mathrm{Re}[X_1 (i\omega Y_1)^\dagger] = \frac{1}{2}\mathrm{Re}[-i\omega X_1 Y_1^\dagger]$$

$$= \frac{1}{2}\mathrm{Re}[-i\omega(Z_\mathrm{R} + iZ_\mathrm{I})] = \frac{1}{2}\omega Z_\mathrm{I} = \frac{1}{2}\omega\mathrm{Im}[X_1 Y_1^\dagger].$$

6.6.2 Cross-sectional mean averaged displacement and pressure of fluid particle

For an inviscid fluid, longitudinal displacement oscillations are not uniform over the cross section of the flow channel because of the viscosity, as we have seen in Chapter 3. Hence, we consider the cross-sectional mean $\langle \xi' \rangle_r$ of displacement oscillation of a given gas parcel, which is also expressed as

$$\langle \xi' \rangle_r = \mathrm{Re}[\xi_{1r} e^{i\omega t}]. \tag{6.40}$$

Also, pressure oscillation is expressed as

$$p' = \mathrm{Re}[p_1 e^{i\omega t}]. \tag{6.41}$$

We set the initial phase of oscillating quantities with reference to $\langle \xi' \rangle_r$. Therefore, $\langle \xi' \rangle_r$ and p' are also expressed as

$$\langle \xi' \rangle_r = |\xi_{1r}| \cos \omega t \tag{6.42}$$

$$p' = |p_1| \cos(\omega t + \theta). \tag{6.43}$$

Hence, a product $p_1 \xi_{1r}^{\dagger}$ is decomposed into the real and imaginary parts as

$$p_1 \xi_{1r}^{\dagger} = |p_1||\xi_{1r}| \cos \theta + i|p_1||\xi_{1r}| \sin \theta.$$

The cross-sectional mean velocity $\langle u' \rangle_r$ is expressed as

$$\langle u' \rangle_r = \mathrm{Re}[u_{1r} e^{i\omega t}] = |u_{1r}| \cos(\omega t + \phi),$$

where $u_{1r} = i\omega \xi_{1r}$ and $\phi = \pi/2 - \theta$. Figure 6.13 presents the phasor representation of ξ_{1r}, p_1, and u_{1r}.

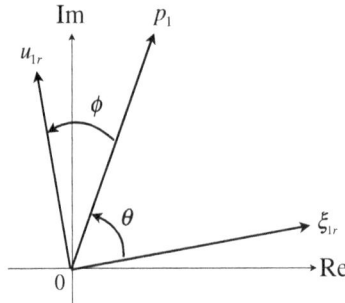

Figure 6.13: Phasor representation of ξ_{1r}, p_1, and u_{1r}. Phase θ represents the phase lead of p_1 relative to ξ_{1r}, whereas ϕ represents the phase lead of u_{1r} relative to p_1.

6.6.3   *Work flux density for viscous fluid*

Work flux density I is expressed as

$$I = \langle\langle p'u'\rangle\rangle.$$

Because p' is uniform over the cross-section of the flow channel, p' is independent of r. Therefore, $\langle\langle p'u'\rangle\rangle = \langle p'\langle u'\rangle_r\rangle_t$. Equation (6.39) yields

$$\langle p'\langle u'\rangle_r\rangle_t = \frac{1}{2}\omega\text{Im}[p_1\xi_{1r}^\dagger],$$

and hence we have

$$I = \frac{\omega}{2}|p_1||\xi_{1r}|\sin\theta, \tag{6.44}$$

or alternatively,

$$I = \frac{1}{2}|p_1||u_{1r}|\cos\phi. \tag{6.45}$$

The expression of I is the same as both for inviscid fluid and viscous fluid.

6.6.4   *Heat flux density for viscous fluid*

Heat flux density Q is expressed as

$$Q = \rho_m T_m \langle\langle S'u'\rangle\rangle.$$

Since $u' = d\xi'/dt$, it is transformed as

$$Q = \rho_m T_m \left\langle\left\langle S'\frac{d\xi'}{dt}\right\rangle\right\rangle.$$

Use of Eq. (6.39) leads to

$$Q = \frac{\omega}{2}\rho_m T_m \langle\text{Im}[S_1\xi_1^\dagger]\rangle_r.$$

The cross-sectional mean of the imaginary part of $S_1\xi_1^\dagger$ is equal to the imaginary part of the cross-sectional mean of it:

$$Q = \frac{\omega}{2}\rho_m T_m \text{Im}[\langle S_1\xi_1^\dagger\rangle_r].$$

By substituting the Lagrangian entropy oscillations (see Section 5.6)

$$S_1 = \left(\frac{\partial S}{\partial p}\right)_T f_\alpha p_1 + \left(\frac{\partial S}{\partial T}\right)_p \frac{dT_m}{dx} b\xi_{1r}$$

we have

$$\langle S_1 \xi_1^\dagger \rangle_r = \left(\frac{\partial S}{\partial p}\right)_T \langle f_\alpha p_1 \xi_1^\dagger \rangle_r + \left(\frac{\partial S}{\partial T}\right)_p \frac{dT_m}{dx} \langle b\xi_{1r}\xi_1^\dagger \rangle_r, \qquad (6.46)$$

where b is defined in Eq. (5.37) by

$$b = \frac{f_\alpha - f_\nu}{(1 - \chi_\nu)(1 - \sigma)}.$$

The solution of the equation of motion for viscous fluid [Eq. (3.82)] is written as

$$\xi_1 = \frac{1 - f_\nu}{1 - \chi_\nu} \xi_{1r}.$$

The first and second term of the right-hand side of Eq. (6.46) are therefore,

$$\left(\frac{\partial S}{\partial p}\right)_T \langle f_\alpha p_1 \xi_1^\dagger \rangle_r = \left(\frac{\partial S}{\partial p}\right)_T \left\langle f_\alpha \frac{1 - f_\nu^\dagger}{1 - \chi_\nu^\dagger} p_1 \xi_{1r}^\dagger \right\rangle_r$$

and

$$\left(\frac{\partial S}{\partial T}\right)_p \frac{dT_m}{dx} \langle b\xi_{1r}\xi_1^\dagger \rangle_r = \left(\frac{\partial S}{\partial T}\right)_p \frac{dT_m}{dx} \left\langle b\frac{1 - f_\nu^\dagger}{1 - \chi_\nu^\dagger} \xi_{1r}\xi_{1r}^\dagger \right\rangle_r$$

$$= \left(\frac{\partial S}{\partial T}\right)_p \frac{dT_m}{dx} |\xi_{1r}|^2 \left\langle b\frac{1 - f_\nu^\dagger}{1 - \chi_\nu^\dagger} \right\rangle_r.$$

Thus, we have

$$\mathrm{Im}[\langle S_1 \xi_1^\dagger \rangle_r] = \left(\frac{\partial S}{\partial p}\right)_T \mathrm{Im}\left[\left\langle f_\alpha \frac{1 - f_\nu^\dagger}{1 - \chi_\nu^\dagger} p_1 \xi_{1r}^\dagger \right\rangle_r\right]$$

$$+ \left(\frac{\partial S}{\partial T}\right)_p \frac{dT_m}{dx} |\xi_{1r}|^2 \mathrm{Im}\left[\left\langle b\frac{1 - f_\nu^\dagger}{1 - \chi_\nu^\dagger} \right\rangle_r\right]. \qquad (6.47)$$

For brevity, we introduce g and g_D which respectively satisfy the following equations

$$g = \left\langle f_\alpha \frac{1 - f_\nu^\dagger}{1 - \chi_\nu^\dagger} \right\rangle_r \qquad (6.48)$$

and

$$\mathrm{Im}\left[\left\langle b \frac{1 - f_\nu^\dagger}{1 - \chi_\nu^\dagger} \right\rangle_r \right] = \mathrm{Im}[g_D]\mathrm{Re}\left[\frac{1}{1 - \chi_\nu} \right]. \qquad (6.49)$$

The first and second terms of Eq. (6.47) are rewritten as

$$\left(\frac{\partial S}{\partial p} \right)_T \mathrm{Im}\left[\left\langle f_\alpha \frac{1 - f_\nu^\dagger}{1 - \chi_\nu^\dagger} p_1 \xi_{1r}^\dagger \right\rangle_r \right]$$

$$= \left(\frac{\partial S}{\partial p} \right)_T |p_1||\xi_{1r}|\{\mathrm{Re}[g]\sin\theta + \mathrm{Im}[g]\cos\theta\}$$

and

$$\left(\frac{\partial S}{\partial T} \right)_p \frac{dT_m}{dx} |\xi_{1r}|^2 \mathrm{Im}\left[\left\langle b \frac{1 - f_\nu^\dagger}{1 - \chi_\nu^\dagger} \right\rangle_r \right]$$

$$= \left(\frac{\partial S}{\partial T} \right)_p \frac{dT_m}{dx} \mathrm{Im}[g_D]\mathrm{Re}\left[\frac{1}{1 - \chi_\nu} \right] |\xi_{1r}|^2.$$

It should be noted that g and g_D are more explicitly expressed by using mathematical formula of Bessel functions (see Appendix) as

$$g = \frac{\chi_\alpha - \chi_\nu^\dagger}{(1 + \sigma)(1 - \chi_\nu^\dagger)} \qquad (6.50)$$

and

$$g_D = \frac{\chi_\alpha - \chi_\nu^\dagger - (1 + \sigma)\chi_\nu + (1 + \sigma)\mathrm{Re}[\chi_\nu]}{(1 - \mathrm{Re}[\chi_\nu])(1 - \sigma^2)}. \qquad (6.51)$$

Now we are able to show the heat flux components. By using the thermodynamic relations of thermal expansion coefficient β in

Eq. (6.13) and of isobaric specific heat in Eq. (6.19), heat flux density Q is decomposed into three components as shown below:

$$Q = Q_{prog} + Q_{stand} + Q_D \tag{6.52}$$

$$Q_{prog} = -\frac{\omega}{2}\beta T_m \text{Re}[g]|p_1||\xi_{1r}|\sin\theta \tag{6.53}$$

$$Q_{stand} = -\frac{\omega}{2}\beta T_m \text{Im}[g]|p_1||\xi_{1r}|\cos\theta \tag{6.54}$$

$$Q_D = \frac{\omega}{2}\rho_m C_p \frac{dT_m}{dx}\text{Im}[g_D]\text{Re}\left[\frac{1}{1-\chi_\nu}\right]|\xi_{1r}|^2. \tag{6.55}$$

If we use $u_{1r} = i\omega\xi_{1r}$, three components are expressed as

$$Q_{prog} = -\frac{1}{2}\beta T_m \text{Re}[g]|p_1||u_{1r}|\cos\phi \tag{6.56}$$

$$Q_{stand} = -\frac{1}{2}\beta T_m \text{Im}[g]|p_1||u_{1r}|\sin\phi \tag{6.57}$$

$$Q_D = \frac{1}{2\omega}\rho_m C_p \frac{dT_m}{dx}\text{Im}[g_D]\text{Re}\left[\frac{1}{1-\chi_\nu}\right]|u_{1r}|^2. \tag{6.58}$$

When Q_{prog}, Q_{stand}, and Q_D for viscous fluid are compared with those for inviscid fluid, we notice the following replacements

$$\chi'_\alpha \leftrightarrow \text{Re}[g], \quad \chi''_\alpha \leftrightarrow \text{Im}[g] \tag{6.59}$$

in Q_{prog} and Q_{stand}, and also

$$\chi''_\alpha \leftrightarrow \text{Im}[g_D]\text{Re}\left[\frac{1}{1-\chi_\nu}\right] \tag{6.60}$$

in Q_D, as well as $\xi_1 \leftrightarrow \xi_{1r}$ and $u_1 \leftrightarrow u_{1r}$. Hence, it would be useful to compare them by graphs shown in Figs. 6.14 and 6.15. When $\sigma = 0.66$ as in the case of He, it is found that $\text{Re}[g] < \chi'_\alpha$, $\text{Im}[g] < \chi''_\alpha$, and $\text{Im}[g_D]\text{Re}[1/(1-\chi_\nu)] < \chi'_\alpha$. However, they are similar to each other when viewed as functions of $\omega\tau_\alpha$. Therefore, the physics behind Q_{prog}, Q_{stand}, and Q_D for the viscous fluids is qualitatively given by illustrative description for the inviscid fluid.

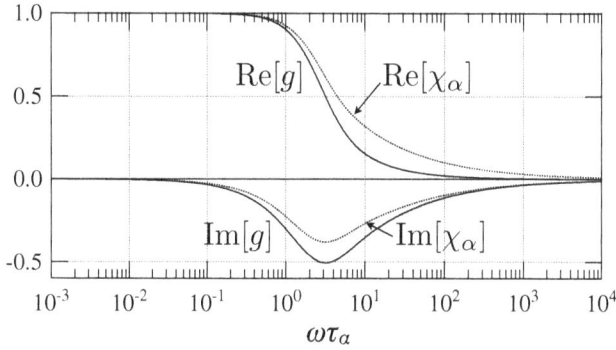

Figure 6.14: $\omega\tau_\alpha$ dependence of g and χ_α.

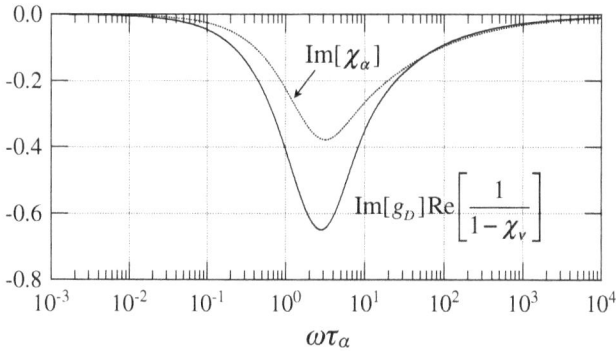

Figure 6.15: $\omega\tau_\alpha$ dependence of $\text{Im}[\chi_\alpha]$ and $\text{Im}[g_D]\text{Re}[\frac{1}{1-\chi_\nu}]$.

6.6.5 Work source for viscous fluid

Work source is expressed for viscous fluid by

$$w = W_\nu + \rho_m \left(\frac{\partial V}{\partial S}\right)_p \left\langle p' \frac{d\langle S'\rangle_r}{dt}\right\rangle_t$$

as we have seen in Section 5.3. The first term W_ν

$$W_\nu = \langle\langle(\mu\Delta u')u'\rangle\rangle$$

on the right-hand side represents the energy dissipation per unit volume and time, as suggested by the subscript ν that represent the kinematic viscosity of the fluid. Inserting the equation of motion for the viscous fluid [see Eq. (3.65)] gives

$$\langle\langle(\mu\Delta u')u'\rangle\rangle = \left\langle\left\langle \frac{\partial p'}{\partial x}u' \right\rangle\right\rangle.$$

By taking the time average, we have

$$W_\nu = \frac{1}{2}\left\langle \mathrm{Re}\left[\frac{dp_1}{dx}u_1^\dagger\right]\right\rangle_r,$$

which is transformed as

$$W_\nu = \frac{1}{2}\mathrm{Re}\left[\left\langle \frac{dp_1}{dx}u_1^\dagger\right\rangle_r\right].$$

Since p_1 is uniform over the cross-section, it is not involved in the cross-sectional mean. Hence

$$W_\nu = \frac{1}{2}\mathrm{Re}\left[\frac{dp_1}{dx}u_{1r}^\dagger\right].$$

The cross-sectional mean velocity u_{1r} is given by a solution of equation of motion for viscous fluid [Eq. (3.80)] as

$$u_{1r} = \frac{i}{\omega\rho_m}\frac{dp_1}{dx}(1 - \chi_\nu),$$

from which we have

$$\frac{dp_1}{dx} = \frac{-i\omega\rho_m}{1 - \chi_\nu}u_{1r}.$$

Therefore, W_ν is further transformed as

$$W_\nu = \frac{1}{2}\mathrm{Re}\left[\frac{-i\omega\rho_m}{1 - \chi_\nu}u_{1r}u_{1r}^\dagger\right] = \frac{1}{2}\omega\rho_m|u_{1r}|^2\mathrm{Re}\left[\frac{-i}{1 - \chi_\nu}\right].$$

Finally we arrive at

$$W_\nu = \frac{1}{2}\omega\rho_m\mathrm{Im}\left[\frac{1}{1 - \chi_\nu}\right]|u_{1r}|^2. \tag{6.61}$$

The second term of w is transformed by taking the temporal mean via Eq. (6.39) as

$$\rho_m \left(\frac{\partial V}{\partial S}\right)_p \left\langle p' \frac{d\langle S'\rangle_r}{dt} \right\rangle_t = \frac{\omega \rho_m}{2} \left(\frac{\partial V}{\partial S}\right)_p \operatorname{Im}[p_1 \langle S_1 \rangle_r^\dagger].$$

Inserting the Lagrangian entropy obtained by the solution of general equation of heat transfer [Eq. (5.38)]

$$\langle S_1 \rangle_r = \left(\frac{\partial S}{\partial p}\right)_T \chi_\alpha p_1 + \left(\frac{\partial S}{\partial T}\right)_p \frac{dT_m}{dx} \langle b \rangle_r \xi_{1r}$$

yields

$$\frac{\omega \rho_m}{2} \left(\frac{\partial V}{\partial S}\right)_p \operatorname{Im}[p_1 \langle S_1 \rangle_r^\dagger]$$

$$= \frac{\omega \rho_m}{2} \left(\frac{\partial V}{\partial S}\right)_p \operatorname{Im}\left[p_1 \left(\frac{\partial S}{\partial p}\right)_T \chi_\alpha^\dagger p_1^\dagger \right]$$

$$+ \frac{\omega \rho_m}{2} \left(\frac{\partial V}{\partial S}\right)_p \operatorname{Im}\left[p_1 \left(\frac{\partial S}{\partial T}\right)_p \frac{dT_m}{dx} \langle b \rangle_r^\dagger \xi_{1r}^\dagger \right], \qquad (6.62)$$

where $\langle b \rangle_r$ is given by [Eq. (5.39)]

$$\langle b \rangle_r = \frac{\chi_\alpha - \chi_\nu}{(1 - \chi_\nu)(1 - \sigma)}.$$

The first term of the right-hand side of Eq. (6.62) is transformed as

$$\frac{\omega \rho_m}{2} \left(\frac{\partial V}{\partial S}\right)_p \operatorname{Im}\left[p_1 \left(\frac{\partial S}{\partial p}\right)_T \chi_\alpha^\dagger p_1^\dagger \right]$$

$$= \frac{\omega \rho_m}{2} \left(\frac{\partial V}{\partial S}\right)_p \left(\frac{\partial S}{\partial p}\right)_T |p_1|^2 \operatorname{Im}[\chi_\alpha^\dagger].$$

By using the thermodynamic relations of Eqs. (6.13) and (6.26), and also the relation of $\text{Im}[\chi_\alpha^\dagger] = -\text{Im}[\chi_\alpha] = -\chi_\alpha''$, we obtain

$$\frac{\omega \rho_m}{2} \left(\frac{\partial V}{\partial S}\right)_p \left(\frac{\partial S}{\partial p}\right)_T |p_1|^2 \text{Im}[\chi_\alpha^\dagger] = \frac{\omega}{2}(K_T - K_S)\chi_\alpha'' |p_1|^2,$$

which represents the energy dissipation due to the thermal conductivity per unit volume and time

$$W_p = \frac{\omega}{2}(K_T - K_S)\chi_\alpha'' |p_1|^2. \tag{6.63}$$

The first term of the right-hand side of Eq. (6.62) is rewritten as

$$\frac{\omega \rho_m}{2} \left(\frac{\partial V}{\partial S}\right)_p \text{Im}\left[p_1 \left(\frac{\partial S}{\partial T}\right)_p \frac{dT_m}{dx} \langle b \rangle_r^\dagger \xi_{1r}^\dagger\right]$$

$$= \frac{\omega \rho_m}{2} \left(\frac{\partial V}{\partial S}\right)_p \left(\frac{\partial S}{\partial T}\right)_p \frac{dT_m}{dx} \text{Im}\left[p_1 \langle b \rangle_r^\dagger \xi_{1r}^\dagger\right].$$

Here, $\text{Im}\left[p_1 \langle b \rangle_r^\dagger \xi_{1r}^\dagger\right]$ is more explicitly written as

$$\text{Im}\left[p_1 \langle b \rangle_r^\dagger \xi_{1r}^\dagger\right] = \text{Im}\left[(\text{Re}\langle b \rangle_r - i\text{Im}\langle b \rangle_r)(|p_1||\xi_{1r}| \cos\theta + i|p_1||\xi_{1r}| \sin\theta)\right]$$

$$= (\text{Re}\langle b \rangle_r)|p_1||\xi_{1r}| \sin\theta - (\text{Im}\langle b \rangle_r)|p_1||\xi_{1r}| \cos\theta.$$

Therefore,

$$\frac{\omega \rho_m}{2} \left(\frac{\partial V}{\partial S}\right)_p \left(\frac{\partial S}{\partial T}\right)_p \frac{dT_m}{dx} \text{Im}\left[p_1 \langle b \rangle_r^\dagger \xi_{1r}^\dagger\right]$$

$$= \frac{\omega}{2}\beta \frac{dT_m}{dx}\{(\text{Re}\langle b \rangle_r)|p_1||\xi_{1r}| \sin\theta - (\text{Im}\langle b \rangle_r)|p_1||\xi_{1r}| \cos\theta\}$$

$$= \frac{\omega}{2}\beta \frac{dT_m}{dx}(\text{Re}\langle b \rangle_r)|p_1||\xi_{1r}| \sin\theta - \frac{\omega}{2}\beta \frac{dT_m}{dx}(\text{Im}\langle b \rangle_r)|p_1||\xi_{1r}| \cos\theta,$$

where the property of partial derivative in Eq. (6.29) was used.

Now we see that the work source w for viscous fluid is expressed by

$$w = W_\nu + W_p + W_{prog} + W_{stand} \qquad (6.64)$$

where

$$W_\nu = \frac{1}{2}\omega\rho_m\mathrm{Im}\left[\frac{1}{1-\chi_\nu}\right]|u_{1r}|^2$$

$$W_p = \frac{\omega}{2}(K_T - K_S)\chi_\alpha''|p_1|^2$$

$$W_{prog} = \frac{\omega}{2}\beta\frac{dT_m}{dx}(\mathrm{Re}\langle b\rangle_r)|p_1||\xi_{1r}|\sin\theta$$

$$W_{stand} = -\frac{\omega}{2}\beta\frac{dT_m}{dx}(\mathrm{Im}\langle b\rangle_r)|p_1||\xi_{1r}|\cos\theta.$$

If we use $u_{1r} = i\omega\xi_{1r}$, W_{prog} and W_{stand} are expressed as

$$W_{prog} = \frac{1}{2}\beta\frac{dT_m}{dx}(\mathrm{Re}\langle b\rangle_r)|p_1||u_{1r}|\cos\phi \qquad (6.65)$$

$$W_{stand} = -\frac{1}{2}\beta\frac{dT_m}{dx}(\mathrm{Im}\langle b\rangle_r)|p_1||u_{1r}|\sin\phi. \qquad (6.66)$$

Let us compare them with those for inviscid fluid. As for W_p, the expression is the same both for viscous and inviscid fluids, whereas W_ν is the work source component that should be considered only for viscous fluid. Figure 6.16 presents a log-log plot of $-\mathrm{Im}[1/(1 - \chi_\nu)]$ as a function of $\omega\tau_\nu$, where we see different slopes for large and small $\omega\tau_\nu$ values. By using the asymptotic forms of χ_ν shown in Section 3.5, those for $-\mathrm{Im}[1/(1 - \chi_\nu)]$ are given by

$$-\mathrm{Im}\left[\frac{1}{1-\chi_\nu}\right] = \begin{cases} \dfrac{4}{\omega\tau_\nu} & (\omega\tau_\nu \ll \pi) \\ \dfrac{1}{\sqrt{\omega\tau_\nu}} & (\omega\tau_\nu \gg \pi). \end{cases}$$

Thus, W_ν is proportional to $\rho_m|u_{1r}|^2/\tau_\nu$ when $\omega\tau_\nu \ll \pi$, whereas it is proportional to $\rho_m|u_{1r}|^2\sqrt{\omega/\tau_\nu}$ when $\omega\tau_\nu \gg \pi$.

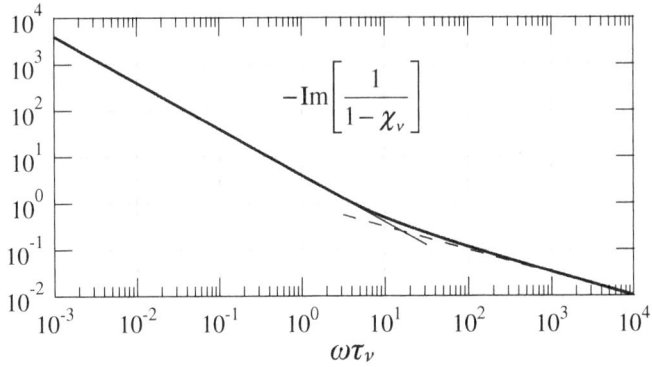

Figure 6.16: $\omega\tau_\nu$ dependence of $-\text{Im}[1/(1-\chi_\nu)]$.

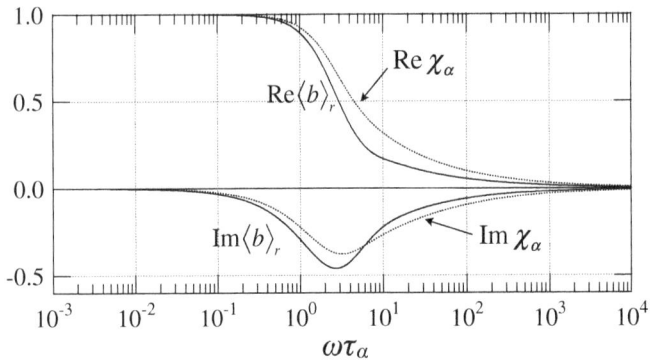

Figure 6.17: $\omega\tau_\alpha$ dependence of χ_α and $\langle b \rangle_r$ when $\sigma = 0.66$.

We notice the following replacement in W_{prog} and W_{stand}

$$\chi_\alpha \leftrightarrow \langle b \rangle_r$$

as well as $\xi_1 \leftrightarrow \xi_{1r}$ and $u_1 \leftrightarrow u_{1r}$. Figure 6.17 shows $\langle b \rangle_r$ as a function of $\omega\tau_\alpha$. As we can see, there is no significant difference between χ_α and $\langle b \rangle_r$, although they coincide with each other only in the limit of $\sigma = 0$. Therefore, the physics behind W_{prog}, W_{stand} for the viscous fluids is qualitatively given by illustrative description given for the inviscid fluid.

6.7 Appendix

6.7.1 *Maxwell relations*

When the internal energy U is considered as a function of entropy S and volume V, its differential dU is expressed as

$$dU = \left(\frac{\partial U}{\partial S}\right)_V dS + \left(\frac{\partial U}{\partial V}\right)_S dV.$$

On the other hand, the first law of thermodynamics states

$$dU = TdS - pdV.$$

Therefore, we see

$$T = \left(\frac{\partial U}{\partial S}\right)_V, \quad -p = \left(\frac{\partial U}{\partial V}\right)_S.$$

The mixed partial derivatives are equal:

$$\frac{\partial}{\partial V}\left(\frac{\partial U}{\partial S}\right) = \frac{\partial}{\partial S}\left(\frac{\partial U}{\partial V}\right).$$

Hence, we obtain

$$\left(\frac{\partial T}{\partial V}\right)_S = -\left(\frac{\partial p}{\partial S}\right)_V. \tag{6.67}$$

The differential of the enthalpy $H = U + pV$ is given as $dH = TdS + Vdp$. Following from this, we have

$$\left(\frac{\partial T}{\partial p}\right)_S = \left(\frac{\partial V}{\partial S}\right)_p. \tag{6.68}$$

In the same way, from the differential of Helmholtz free energy $F = U - TS$ and that of Gibbs free energy $G = U - TS + pV$, we obtain

$$\left(\frac{\partial S}{\partial V}\right)_T = \left(\frac{\partial p}{\partial T}\right)_V, \tag{6.69}$$

and

$$\left(\frac{\partial S}{\partial p}\right)_T = -\left(\frac{\partial V}{\partial T}\right)_p. \tag{6.70}$$

Equations (6.67)–(6.70) are called Maxwell relations.

6.7.2 Properties of partial derivative

For a function $f = f(x, y)$, the (total) differential df is given by

$$df = \left(\frac{\partial f}{\partial x}\right)_y dx + \left(\frac{\partial f}{\partial y}\right)_x dy.$$

When x is considered as functions of y and f, the differential dx is given by

$$dx = \left(\frac{\partial x}{\partial y}\right)_f dy + \left(\frac{\partial x}{\partial f}\right)_y df.$$

Combining these equations gives

$$df = \left(\frac{\partial f}{\partial x}\right)_y \left[\left(\frac{\partial x}{\partial y}\right)_f dy + \left(\frac{\partial x}{\partial f}\right)_y df\right] + \left(\frac{\partial f}{\partial y}\right)_x dy.$$

Since f and y are individual variables, we have the following relations

$$\left(\frac{\partial f}{\partial x}\right)_y \left(\frac{\partial x}{\partial f}\right)_y = 1 \tag{6.71}$$

$$-\left(\frac{\partial f}{\partial x}\right)_y \left(\frac{\partial x}{\partial y}\right)_f = \left(\frac{\partial f}{\partial y}\right)_x. \tag{6.72}$$

In the same way, for a function $g = g(x, y)$, we have the following equation

$$dx = \left(\frac{\partial x}{\partial y}\right)_g dy + \left(\frac{\partial x}{\partial g}\right)_y dg,$$

when x is also considered as functions of y and g. The differential df is then written as

$$df = \left(\frac{\partial f}{\partial x}\right)_y \left[\left(\frac{\partial x}{\partial y}\right)_g dy + \left(\frac{\partial x}{\partial g}\right)_y dg\right] + \left(\frac{\partial f}{\partial y}\right)_x dy.$$

Therefore,

$$df = \left[\left(\frac{\partial f}{\partial x}\right)_y \left(\frac{\partial x}{\partial y}\right)_g + \left(\frac{\partial f}{\partial y}\right)_x\right] dy + \left(\frac{\partial f}{\partial x}\right)_y \left(\frac{\partial x}{\partial g}\right)_y dg$$

$$= \left[\left(\frac{\partial f}{\partial x}\right)_y \left(\frac{\partial x}{\partial y}\right)_g + \left(\frac{\partial f}{\partial y}\right)_x\right] dy + \left(\frac{\partial f}{\partial g}\right)_y dg.$$

If the function f is considered as $f = f(x(g), y) = f(g, y)$, the differential df is expressed by

$$df = \left(\frac{\partial f}{\partial y}\right)_g dy + \left(\frac{\partial f}{\partial g}\right)_y dg.$$

Therefore, we have the relation

$$\left(\frac{\partial f}{\partial y}\right)_g = \left(\frac{\partial f}{\partial y}\right)_x + \left(\frac{\partial f}{\partial x}\right)_y \left(\frac{\partial x}{\partial y}\right)_g. \tag{6.73}$$

6.7.3 *Some useful thermodynamic relations*

Thermal expansion coefficient β, Isothermal compressibility K_T, adiabatic compressibility K_S, isobaric specific heat C_p, and isochoric specific heat C_V are given by

$$\beta = \frac{1}{V}\left(\frac{\partial V}{\partial T}\right)_p, \quad K_T = -\frac{1}{V}\left(\frac{\partial V}{\partial p}\right)_T, \quad K_S = -\frac{1}{V}\left(\frac{\partial V}{\partial p}\right)_S,$$

$$C_p = T\left(\frac{\partial S}{\partial T}\right)_p, \quad C_V = T\left(\frac{\partial S}{\partial T}\right)_V.$$

Adiabatic compressibility K_S is linked with adiabatic speed of sound, $c_S = \sqrt{(\partial p/\partial \rho)_S}$, and density $\rho = 1/V$ as

$$\rho K_S c_S^2 = 1. \tag{6.74}$$

The Maxwell relation in Eq. (6.70)

$$\left(\frac{\partial S}{\partial p}\right)_T = -\left(\frac{\partial V}{\partial T}\right)_p,$$

is rewritten by using β as

$$\left(\frac{\partial S}{\partial p}\right)_T = -\frac{\beta}{\rho},$$

which confirms the thermodynamic relation in Eq. (6.13).

Partial derivative in Eq. 6.26 is transformed through Eq. (6.73) as

$$\left(\frac{\partial V}{\partial S}\right)_p = \left(\frac{\partial V}{\partial S}\right)_T + \left(\frac{\partial V}{\partial T}\right)_S \left(\frac{\partial T}{\partial S}\right)_p,$$

which can be further transformed as

$$\left(\frac{\partial V}{\partial S}\right)_p = \left(\frac{\partial V}{\partial p}\right)_T \left(\frac{\partial p}{\partial S}\right)_T + \left(\frac{\partial V}{\partial p}\right)_S \left(\frac{\partial p}{\partial T}\right)_S \left(\frac{\partial T}{\partial S}\right)_p.$$

If we use Eq. (6.72), we have

$$\left(\frac{\partial p}{\partial T}\right)_S \left(\frac{\partial T}{\partial S}\right)_p = -\left(\frac{\partial p}{\partial S}\right)_T.$$

Therefore,

$$\left(\frac{\partial V}{\partial S}\right)_p = \left(\frac{\partial p}{\partial S}\right)_T \left[\left(\frac{\partial V}{\partial p}\right)_T - \left(\frac{\partial V}{\partial p}\right)_S\right],$$

from which we obtain

$$\left(\frac{\partial V}{\partial S}\right)_p = \frac{K_T - K_S}{\beta} \tag{6.75}$$

in Eq. (6.26).

Partial derivative $(\partial V/\partial S)_p$ is expressed also as

$$\left(\frac{\partial V}{\partial S}\right)_p = \left(\frac{\partial V}{\partial T}\right)_p \left(\frac{\partial T}{\partial S}\right)_p = \beta V \frac{T}{C_p} = \frac{\beta T}{\rho C_p}.$$

Comparison with equation above yields

$$\frac{K_T - K_S}{\beta} = \frac{\beta T}{\rho C_p}, \tag{6.76}$$

which will be used in Chapter 8.

The ratio of K_T to K_S is expressed by

$$\frac{K_T}{K_S} = \frac{\left(\frac{\partial V}{\partial p}\right)_T}{\left(\frac{\partial V}{\partial p}\right)_S} = \frac{\left(\frac{\partial V}{\partial T}\right)_p \left(\frac{\partial T}{\partial p}\right)_V}{\left(\frac{\partial V}{\partial S}\right)_p \left(\frac{\partial S}{\partial p}\right)_V}.$$

Because

$$\left(\frac{\partial V}{\partial T}\right)_p = \left(\frac{\partial V}{\partial S}\right)_p \left(\frac{\partial S}{\partial T}\right)_p, \quad \left(\frac{\partial S}{\partial p}\right)_V = \left(\frac{\partial S}{\partial T}\right)_V \left(\frac{\partial T}{\partial p}\right)_V,$$

the ratio K_T/K_S is reduced to

$$\frac{K_T}{K_S} = \frac{\left(\frac{\partial S}{\partial T}\right)_p}{\left(\frac{\partial S}{\partial T}\right)_V},$$

in other words,

$$\frac{K_T}{K_S} = \frac{C_p}{C_V} = \gamma, \tag{6.77}$$

where γ is the specific heat ratio. Inserting Eq. (6.77) into Eq. (6.76) gives

$$\frac{K_S}{\beta}(\gamma - 1) = \frac{\beta T}{\rho C_p},$$

which can be transformed as

$$\frac{\beta T}{\rho C_p} = \frac{\gamma - 1}{\beta \rho c_S^2}$$

by using Eq. (6.74). In an ideal gas, the relation $\beta T = 1$ holds. Therefore,

$$\rho C_p = \frac{\rho c_S^2}{T(\gamma - 1)}. \tag{6.78}$$

This thermodynamic relation will be used in Chapters 7 and 8.

6.7.4   *Derivation of g and g_D*

Function g is defined by

$$g = \left\langle f_\alpha \frac{1 - f_\nu^\dagger}{1 - \chi_\nu^\dagger} \right\rangle_r = \frac{\chi_\alpha - \langle f_\nu^\dagger f_\alpha \rangle_r}{1 - \chi_\nu^\dagger}. \tag{6.79}$$

The term $\langle f_\nu^\dagger f_\alpha \rangle_r$ is written as

$$\langle f_\nu^\dagger f_\alpha \rangle_r = \frac{1}{\pi r_0^2} \int_0^{r_0} 2\pi r \frac{J_0(\eta_\nu^\dagger)}{J_0(\eta_{\nu 0}^\dagger)} \frac{J_0(\eta_\alpha)}{J_0(\eta_{\alpha 0})} \, dr$$

$$= \frac{2}{\eta_{\nu 0}^2} \int_0^{\eta_{\nu 0}} \frac{J_0(i\eta_\nu)}{J_0(i\eta_{\nu 0})} \frac{J_0(\sqrt{\sigma}\eta_\nu)}{J_0(\sqrt{\sigma}\eta_{\nu 0})} \, d\eta_\nu$$

$$= \frac{2}{\eta_{\nu 0}^2} \frac{1}{J_0(i\eta_{\nu 0}) J_0(\eta_{\alpha 0})} \int_0^{\eta_{\nu 0}} J_0(i\eta_\nu) J_0(\sqrt{\sigma}\eta_\nu) \, d\eta_\nu.$$

From the formula of the Bessel functions, we have

$$\int_0^{z_0} J_0(Az) J_0(Bz) \, dz = \frac{z_0}{A^2 - B^2} \left[A J_1(Az) J_0(Bz) - B J_0(Az) J_1(Bz) \right]. \tag{6.80}$$

Therefore,

$$\int_0^{\eta_{\nu 0}} J_0(i\eta_\nu) J_0(\sqrt{\sigma}\eta_\nu) \, d\eta_\nu = \frac{1}{1+\sigma} \left[\frac{\eta_{\nu 0} J_1(i\eta_{\nu 0}) J_0(\eta_{\alpha 0})}{i} \right.$$

$$\left. - \sqrt{\sigma} J_0(i\eta_{\nu 0}) J_1(\eta_{\alpha 0})) \right],$$

and hence we have

$$\langle f_\nu^\dagger f_\alpha \rangle_r = \frac{1}{1+\sigma} \left[\frac{2 J_1(i\eta_{\nu 0})}{i\eta_{\nu 0} J_0(i\eta_{\nu 0})} + \sigma \frac{2 J_1(\eta_{\alpha 0})}{\eta_{\alpha 0} J_0(\eta_{\alpha 0})} \right]$$

$$= \frac{\chi_\nu^\dagger + \sigma \chi_\alpha}{1+\sigma}.$$

As a result we obtain

$$g = \frac{\chi_\alpha - \chi_\nu^\dagger}{(1+\sigma)(1 - \chi_\nu^\dagger)}.$$

Function g_D satisfies

$$\mathrm{Im}[g_D] = \frac{\mathrm{Im}\left[\left\langle b\frac{1 - f_\nu^\dagger}{1 - \chi_\nu^\dagger}\right\rangle_r\right]}{\mathrm{Re}\left[\frac{1}{1 - \chi_\nu}\right]},$$

where

$$b = \frac{f_\alpha - f_\nu}{(1 - \chi_\nu)(1 - \sigma)}.$$

Thus, we are able to define

$$g_D = \frac{\left\langle b\frac{1 - f_\nu^\dagger}{1 - \chi_\nu^\dagger}\right\rangle_r}{\mathrm{Re}\left[\frac{1}{1 - \chi_\nu}\right]} = \frac{\chi_\alpha - \chi_\nu - \langle f_\nu^\dagger f_\alpha\rangle_r + \langle f_\nu^\dagger f_\nu\rangle_r}{(1 - \sigma)\mathrm{Re}\left[\frac{1}{1 - \chi_\nu}\right]|1 - \chi_\nu|^2}.$$

From the discussion above, we see

$$\langle f_\nu^\dagger f_\alpha\rangle_r = \frac{\chi_\nu^\dagger + \sigma\chi_\alpha}{1 + \sigma}, \quad \langle f_\nu^\dagger f_\nu\rangle_r = \frac{\chi_\nu^\dagger + \chi_\nu}{2} = \mathrm{Re}[\chi_\nu].$$

Also, we see the relation

$$\mathrm{Re}\left[\frac{1}{1 - \chi_\nu}\right]|1 - \chi_\nu|^2 = 1 - \mathrm{Re}[\chi_\nu].$$

Therefore, we obtain

$$g_D = \frac{\chi_\alpha - \chi_\nu^\dagger - (1 + \sigma)\chi_\nu + (1 + \sigma)\mathrm{Re}[\chi_\nu]}{(1 - \mathrm{Re}[\chi_\nu])(1 - \sigma^2)}.$$

6.7.5   *Expression of Q_D*

As shown in Section 6.6.4, Q_D is given by

$$Q_D = \frac{1}{2\omega}\rho_m C_p\frac{dT_m}{dx}\mathrm{Im}[g_D]\mathrm{Re}\left[\frac{1}{1 - \chi_\nu}\right]|u_{1r}|^2, \qquad (6.81)$$

where

$$g_D = \frac{\chi_\alpha - \chi_\nu^\dagger - (1+\sigma)\chi_\nu + (1+\sigma)\mathrm{Re}[\chi_\nu]}{(1 - \mathrm{Re}[\chi_\nu])(1-\sigma^2)}.$$

This expression of Q_D followed Tominaga's text, and it is equivalent to that given in Swift's text. The explanation is given below.

When χ_ν is expressed as $\chi_\nu = \chi_\nu' + i\chi_\nu''$,

$$\mathrm{Re}\left[\frac{1}{1-\chi_\nu}\right] = \mathrm{Re}\left[\frac{1}{1-\chi_\nu' - i\chi_\nu''}\right] = \mathrm{Re}\left[\frac{1-\chi_\nu' + i\chi_\nu''}{(1-\chi_\nu')^2 + \chi_\nu''^2}\right]$$

$$= \frac{1-\chi_\nu'}{(1-\chi_\nu')^2 + \chi_\nu''^2}.$$

Since $|1-\chi_\nu|^2 = (1-\chi_\nu')^2 + \chi_\nu''^2$,

$$\mathrm{Re}\left[\frac{1}{1-\chi_\nu}\right] = \frac{1-\chi_\nu'}{|1-\chi_\nu|^2}.$$

Use of $\chi_\nu = \chi_\nu' + i\chi_\nu''$ transforms g_D as

$$g_D = \frac{\chi_\alpha - (\chi_\nu' - i\chi_\nu'') - (1+\sigma)(\chi_\nu' + i\chi_\nu'') + (1+\sigma)\chi_\nu'}{(1-\chi_\nu')(1-\sigma^2)}$$

$$= \frac{\chi_\alpha' - \chi_\nu' + i(\chi_\alpha'' - \sigma\chi_\nu'')}{(1-\chi_\nu')(1-\sigma^2)}.$$

Therefore, $\mathrm{Im}[g_D]$ is given by

$$\mathrm{Im}[g_D] = \frac{\chi_\alpha'' - \sigma\chi_\nu''}{(1-\chi_\nu')(1-\sigma^2)} = \frac{\mathrm{Im}[\chi_\alpha + \sigma\chi_\nu^\dagger]}{(1-\chi_\nu')(1-\sigma^2)}.$$

Now we are able to rewrite Q_D as

$$Q_D = \frac{1}{2\omega}\rho_m C_p \frac{dT_m}{dx} \frac{\mathrm{Im}[\chi_\alpha + \sigma\chi_\nu^\dagger]}{(1-\chi_\nu')(1-\sigma^2)} \frac{1-\chi_\nu'}{|1-\chi_\nu|^2}|u_{1r}|^2,$$

and hence we obtain

$$Q_D = \frac{\rho_m C_p|u_{1r}|^2}{2\omega(1-\sigma^2)|1-\chi_\nu|^2}\mathrm{Im}\left[\chi_\alpha + \sigma\chi_\nu^\dagger\right]\frac{dT_m}{dx},$$

which is the expression given in the Swift's text as the enthalpy flow component proportional to dT_m/dx.

6.8 Problems

1 Suppose that displacement oscillation and pressure oscillation are expressed by $\xi' = |\xi_1|\cos(\omega t)$ and $p' = |p_1|\cos(\omega t+\theta)$, respectively. Draw schematically the $p' - \xi'$ diagrams when $\theta = 0, \pi/2$ and $-\pi/2$.

2 Suppose that the cross-sectional mean entropy oscillation is given by

$$\langle S_1 \rangle_r = \chi_\alpha \left(\frac{\partial S}{\partial p} \right)_T p_1, \quad \left(\frac{\partial S}{\partial p} \right)_T < 0.$$

(1) Draw schematically the phasor diagrams of p_1 and $\langle S_1 \rangle_r$, when $\omega \tau_\alpha \ll 1$, $\omega \tau_\alpha \sim 1$, and $\omega \tau_\alpha \gg 1$.
(2) Draw schematically the $p' - \langle S' \rangle_r$ diagrams when $\omega \tau_\alpha \ll 1$, $\omega \tau_\alpha \sim 1$, and $\omega \tau_\alpha \gg 1$.

3 Suppose that the cross-sectional mean entropy oscillation is given by

$$\langle S_1 \rangle_r = \chi_\alpha \frac{C_p}{T_m} \frac{dT_m}{dx} \xi_1, \quad \frac{dT_m}{dx} > 0.$$

(1) Draw schematically the phasor diagrams of ξ_1 and $\langle S_1 \rangle_r$, when $\omega \tau_\alpha \ll 1$, $\omega \tau_\alpha \sim 1$, and $\omega \tau_\alpha \gg 1$.
(2) Draw schematically the $\langle S' \rangle_r - \xi'$ diagrams when $\omega \tau_\alpha \ll 1$, $\omega \tau_\alpha \sim 1$, and $\omega \tau_\alpha \gg 1$.

Answers

1

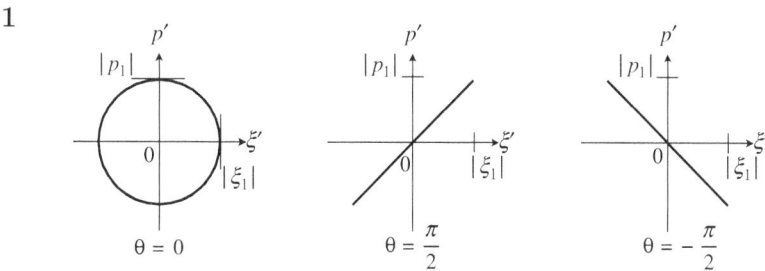

Figure 6.18: $p' - \xi'$ diagrams when $\theta = 0$, $\pi/2$, and $-\pi/2$.

2 See Fig. 6.19.
3 See Fig. 6.20.

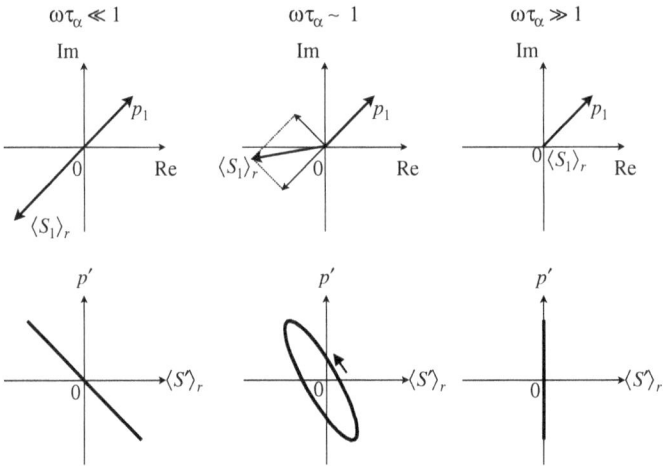

Figure 6.19: Phasor diagrams of p_1 and $\langle S_1 \rangle_r$, when $\omega\tau_\alpha \ll 1$, $\omega\tau_\alpha \sim 1$, and $\omega\tau_\alpha \gg 1$ (upper row) and the corresponding $p' - \langle S' \rangle_r$ diagrams (lower row).

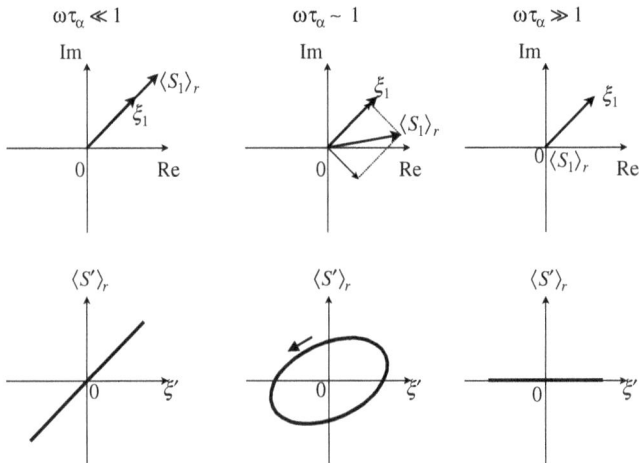

Figure 6.20: Phasor diagrams of ξ_1 and $\langle S_1 \rangle_r$, when $\omega\tau_\alpha \ll 1$, $\omega\tau_\alpha \sim 1$, and $\omega\tau_\alpha \gg 1$ (upper row) and the corresponding $\langle S' \rangle_r - \xi'$ diagrams (lower row).

Chapter 7

Work Source

The work source caused by oscillating fluid parcels has four components: W_{prog} due to traveling wave component of pressure oscillation; W_{stand} due to standing wave component of pressure oscillation; W_p and W_ν due to thermal conductivity and viscosity of the fluid. The work source components W_{prog} and W_{stand} can contribute favorably to acoustic heat engines to produce acoustic power, whereas W_p and W_ν are always associated with energy dissipation. In this chapter, thermoacoustic self-sustained oscillations and acoustic heat engines are discussed in terms of these work source components to gain a deeper understanding of them.

7.1 Acoustic Power Amplification by Temperature Gradients

7.1.1 Ceperley's proposal

Ceperley's proposal on pistonless Stirling engine [31, 65, 66] triggered broad interest from researchers on the thermodynamic point of view. The key was a problem of traveling acoustic waves in a pipe filled with working gas when it passes through a regenerator with positive temperature gradient, as shown in Fig. 7.1. Let's review his proposal briefly. Hereafter the subscripts C and H mean the values at cold and hot ends of the physical quantity, respectively, while the prime ($'$) means the oscillating part of it.

   Suppose that the pressure at the cold end, $p_C = p_m + p'_C$, can be seen as the same as that at the hot end, $p_H = p_m + p'_H$,

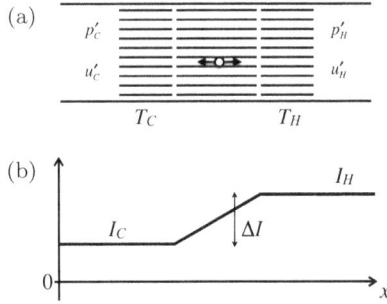

Figure 7.1: Regenerator with a (a) positive temperature gradient and (b) axial distribution of acoustic power along the regenerator. Acoustic power is amplified as it passes through the regenerator. Acoustic power increase ΔI represents the result of thermoacoustic energy conversion from heat flow to work flow (acoustic power).

by assuming the shortness of the regenerator and the low impact of viscous attenuation;

$$p'_C = p'_H. \tag{7.1}$$

Since the mass flow rate is conserved at ends of the regenerator, we have

$$\rho_C U_C = \rho_H U_H, \tag{7.2}$$

where ρ_k $(k = C, H)$ is the temporal mean density of the gas, $U_k = A_k \langle u'_k \rangle_r$ represents the volume velocity given by product of cross-sectional area and cross-sectional mean velocity. When the working gas is assumed by an ideal gas, the state equation gives the relation

$$\frac{\rho_C T_C}{p_C} = \frac{\rho_H T_H}{p_H}. \tag{7.3}$$

Since $p_C = p_H$, we have

$$\frac{U_H}{U_C} = \frac{T_H}{T_C}, \tag{7.4}$$

which indicates that the volume velocity is amplified by the axial temperature gradient by a factor of T_H/T_C. As a result, the work flow $\tilde{I} = A\langle p'\langle u'\rangle_r\rangle_t$ also obeys

$$\frac{\tilde{I}_H}{\tilde{I}_C} = \frac{T_H}{T_C}. \tag{7.5}$$

This relation represents that the regenerator serves as an acoustic power amplifier of a gain T_H/T_C.

The acoustic power amplification is achieved by the acoustic power production

$$\Delta\tilde{I} = \tilde{I}_H - \tilde{I}_C \tag{7.6}$$

resulting from the thermoacoustic energy conversion from heat flow to work flow in the regenerator. The thermodynamic cycle that the gas parcels execute in the regenerator can be illustrated by plotting pressure oscillation against displacement oscillation, as presented in Fig. 7.2(a). When the traveling wave goes in a positive direction of x, the displacement oscillation of the gas parcel lags from pressure oscillation by $\pi/2$. Thus, an ellipse is drawn in clockwise direction. The thermodynamic processes are roughly considered in four time intervals between A, B, C, D.

In A–B, the pressure increases while the gas parcel stays almost at rest. Because of tiny flow channels in the regenerator, isothermal compression process takes place; in B–C, the gas displacement occurs from cold to hot along the temperature gradient in the regenerator, which is shown in Fig. 7.2(b). Thus, this time interval corresponds to the heating process, in which the gas parcel is heated by the surrounding solid wall. In C–D, the gas parcel experiences isothermal

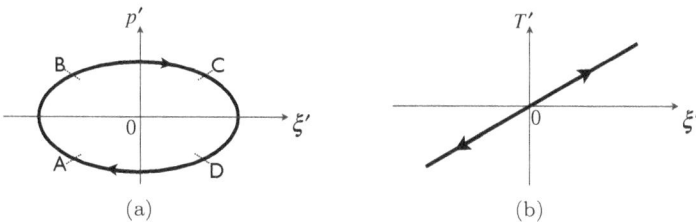

Figure 7.2: Traces of gas particles in the regenerator in (a) pressure versus displacement diagram and (b) temperature versus displacement diagram.

expansion process, whereas in D–A, it undergoes cooling process as it returns from hot to cold. These series of thermodynamic processes constitute the thermodynamic cycle equivalent to the Stirling cycle. More specifically, W_{prog} is responsible for this type of energy conversion process.

Ceperley has tried experimental verification of his proposal, in addition to theoretical analysis [31]. He employed a stainless steel wool based regenerator and placed it between two coil heaters installed in a gas filled pipe. By heating up one of them, a positive or negative temperature gradient was created along the regenerator. The pipe was connected to a loudspeaker at an end, and terminated by end resistance at the other end, to eliminate reflection of transmitted acoustic waves. The gas was air at ambient pressure. The acoustic power entering and leaving the regenerator was measured with and without the temperature gradient, when the loudspeaker emitted a sound of 190 Hz. The acoustic power gain of $G = 0.90$ was obtained with the positive temperature gradient. This gain was greater than $G = 0.81$ recorded without the temperature gradient and $G = 0.70$ with the negative temperature gradient, but still less than unity, meaning that the acoustic power amplification was not achieved.

The problem in his experiment would be related to the difficulty in estimating the effective pore size in the steel wool regenerator. If the pore size was too large, one should make it smaller to achieve the isothermal process; if it was too small, one should make it larger to reduce the viscous damping. However, the effective pore size was difficult to figure out in complicated flow channel of steel wool.

7.1.2   *Experiments in traveling wave field*

Ceperley's proposal was experimentally verified later by more sophisticated experiments using regenerators having regular pores made of honeycomb ceramic catalyst support with many square pores [67]. Figure 7.3 schematically presents the experimental setup where the regenerator was installed in a long tube that was connected

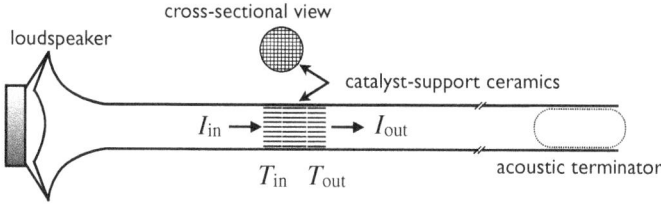

Figure 7.3: Experimental setup to study the acoustic power amplification by temperature gradient along a regenerator.

to a loudspeaker and an acoustic terminator. The regenerator was sandwiched by hot and cold heat exchangers in the way that the sound waves from the loudspeaker go from cold to hot through the regenerator. The alignment of cold heat exchanger and hot heat exchanger was reversed when the negative temperature gradient was tested. The length of regenerator (20 mm) and those of heat exchangers (10 mm) are short compared to the wavelength of the sound waves tested in the experiments.

A parameter governing thermal interactions between the gas and solid walls, $\omega\tau_\alpha$, is a product of angular frequency $\omega = 2\pi f$ and thermal relaxation time $\tau_\alpha = r_0^2/(2\alpha)$, for the gas to thermally equilibrate with the wall, where f is the frequency of sound, r_0 is the channel radius of the regenerator. As we have discussed in Chapter 5, $\omega\tau_\alpha \ll 1$ means isothermal condition for the gas, while $\omega\tau_\alpha \gg 1$ means the adiabatic condition. Another parameter $\omega\tau_\nu$ governs the viscous interaction of the gas with the wall, where $\tau_\nu = r_0^2/(2\nu)$ is the viscous relaxation time (ν is kinematic viscosity of the gas). In Chapter 3, we have shown that $\omega\tau_\nu \ll 1$ gives extremely strong viscous damping, while $\omega\tau_\nu \gg 1$ makes viscous damping negligibly small. For most gases, the Prandtl number $\sigma = \nu/\alpha$ is approximately 0.7 ($\sigma = 0.66$ for helium and 0.71 for air, at ambient pressure and temperature). Therefore, the impacts of thermal conductivity and viscosity are not controlled individually. To control the thermal (and viscous) interactions in a wide range, one can change the frequency f and the channel radius r_0 of the regenerator. In this experiment, three regenerators were employed. Each of them had many square pores with the size $2r_0 \times 2r_0$, but has different r_0. A swept-sine signal

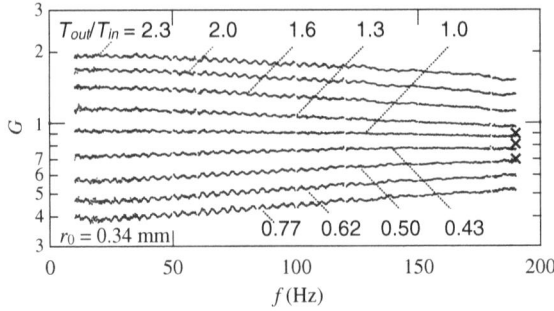

Figure 7.4: Acoustic power gain G as a function of frequency of sound. The channel radius is $r_0 = 0.34$ mm.

changing from 10 Hz to 190 Hz was generated by a synthesizer and fed to a loudspeaker.

Figure 7.4 presents the acoustic power gain $G = I_{out}/I_{in}$ obtained with the regenerator of $r_0 = 0.34$ mm. Temperatures T_{in} and T_{out} in Fig. 7.4 respectively represent the end temperatures of the regenerator at the inlet and the outlet. Hence, when $T_{out}/T_{in} > 1$, the sound waves go from cold to hot. With $T_{out}/T_{in} \geq 1.6$, the acoustic power amplification was achieved in all the frequency range tested as manifested by $G > 1$. With $T_{out}/T_{in} \geq 1.3$, G crosses 1 around $f < 150$ Hz, where the acoustic power production by the thermodynamic cycles of the gas balances with viscous and thermal damping. When $T_{out}/T_{in} < 1.0$, the acoustic power is more strongly reduced than when $T_{out}/T_{in} = 1.0$. This result indicates that the thermodynamic cycles of the gas contributed to attenuation of acoustic power.

The three symbols ($\times$) in the figure represent the experimental gains that Ceperley reported. In his experiment, the temperature ratio was $T_{out}/T_{in} = 1.16, 1.0, 0.86$. If he had increased the temperature ratio of hot to cold part or reduced the sound frequency, he would have successfully observed the acoustic power amplification.

Figure 7.5 summarizes the experimental results obtained with three regenerators of $r_0 = 0.34, 0.56, 0.75$ mm. The horizontal axis is $\omega\tau_\alpha$, so the results with the smaller r_0 are plotted in the lower $\omega\tau_\alpha$ region. The vertical axis presents the difference ΔG given by

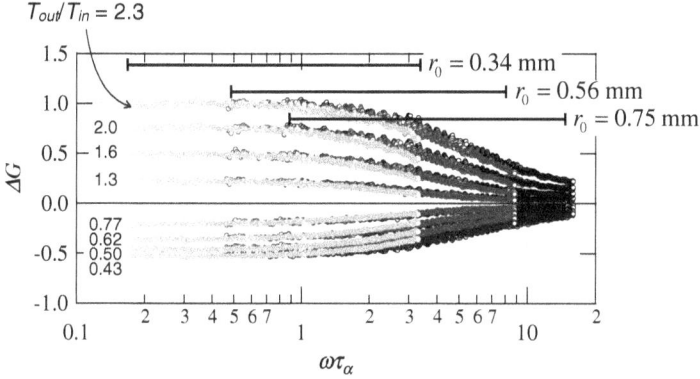

Figure 7.5: Difference of acoustic power gain ΔG between the gains G with and without the temperature gradient as a function of frequency of sound $\omega\tau_\alpha$. The channel radii are $r_0 = 0.34, 0.56, 0.75$ mm.

$\Delta G = G - G_0$, where G_0 means the gain without the temperature gradient in the regenerator. Plotting ΔG was intended to see only the effect of temperature gradients by extracting the effects of thermal and viscous damping. Although the experimental data were obtained with various r_0 and f, ΔG falls onto common curves specified by the temperature ratio. This result demonstrates that $\omega\tau_\alpha$ is the universal parameter that governs the thermoacoustic power amplification.

It should be worth trying to interpret the experimental G in terms of the work source discussed in Chapter 6. We consider the pressure oscillation and cross-sectional mean velocity, whose complex amplitudes are p_1 and u_{1r}. Suppose that the phase of u_{1r} is advanced from p_1 by ϕ, as shown in Fig. 7.6. Since the forward traveling wave is considered, ϕ is limited to $-\pi/2 < \phi < \pi/2$. The regenerator has a uniform cross-section and it is much shorter than the wavelength of the sound. Therefore, we can neglect the axial change of p_1 and u_{1r} in the following discussion.

The work source w in the regenerator with uniform cross-section can be obtained by $w = \Delta I/\Delta x$, where Δ denotes the increase of the work flux density between the ends of the regenerator, whereas Δx signifies the regenerator length. If the sound goes from cold to hot through the regenerator, the work flux density at the hot end I_H is

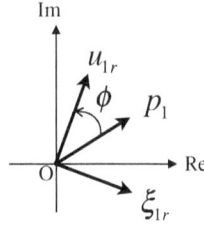

Figure 7.6: Phasor representation of p_1, u_{1r}, and ξ_{1r}. The phase lead of u_{1r} relative to p_1 is denoted by ϕ.

expressed by $I_H = I_C + w\Delta x$. Therefore, the gain is obtained by

$$G = 1 + \frac{w}{I_C}\Delta x. \tag{7.7}$$

The work source w is a sum of four components W_{prog}, W_{stand}, W_p, and W_ν, but in the traveling-wave like field with $\phi \approx 0$, W_{stand} is negligible. Also, in the absence of the temperature gradient, there are no contributions from W_{prog} and W_{stand}. Therefore, G and G_0 are approximated by

$$G = 1 + \frac{W_{prog} + W_p + W_\nu}{I_C}\Delta x \tag{7.8}$$

and

$$G_0 = 1 + \frac{W_p + W_\nu}{I_C}\Delta x. \tag{7.9}$$

So, $\Delta G = G - G_0$ is given by

$$\Delta G = \frac{W_{prog}}{I_C}\Delta x. \tag{7.10}$$

If the temperature distribution is approximated by a linear function of x,

$$I_C \sim I\frac{2T_C}{T_H + T_C} \tag{7.11}$$

where I is the work flux density at the middle of the regenerator. We have assumed here that the work flux density is proportional to the temperature ($I_H/I_C = T_H/T_C$), following Ceperley's discussion. Thus, the gain difference ΔG is given by

$$\Delta G \sim \frac{W_{prog}\Delta x}{I}\frac{T_H + T_C}{2T_C}. \tag{7.12}$$

Since I and W_{prog} are given by

$$I = \frac{1}{2}|p_1||u_{1r}|\cos\phi \tag{7.13}$$

and

$$W_{prog} = \frac{1}{2}\beta\frac{dT_m}{dx}(\mathrm{Re}\langle b\rangle_r)|p_1||u_{1r}|\cos\phi, \tag{7.14}$$

where

$$\langle b\rangle_r = \frac{\chi_\alpha - \chi_\nu}{(1 - \chi_\nu)(1 - \sigma)}. \tag{7.15}$$

Therefore, we have

$$\Delta G \sim \beta\frac{dT_m}{dx}(\mathrm{Re}\langle b\rangle_r)\Delta x\frac{T_H + T_C}{2T_C}, \tag{7.16}$$

which can be transformed as

$$\Delta G \sim \left(\frac{T_H}{T_C} - 1\right)\mathrm{Re}\langle b\rangle_r. \tag{7.17}$$

Here we have used

$$\frac{dT_m}{dx}\Delta x \sim T_H - T_C \tag{7.18}$$

and

$$\beta = \frac{2}{T_H + T_C} \tag{7.19}$$

from the thermodynamic relation $\beta T = 1$ for an ideal gas. Equation (7.17) demonstrates that the ΔG can be essentially represented by the temperature ratio T_H/T_C and $\mathrm{Re}\langle b\rangle_r$. As shown in Fig. 7.7, the $\omega\tau_\alpha$ dependence of $\mathrm{Re}\langle b\rangle_r$ well reproduces a drastic change of ΔG at $\omega\tau_\alpha \approx 3$.

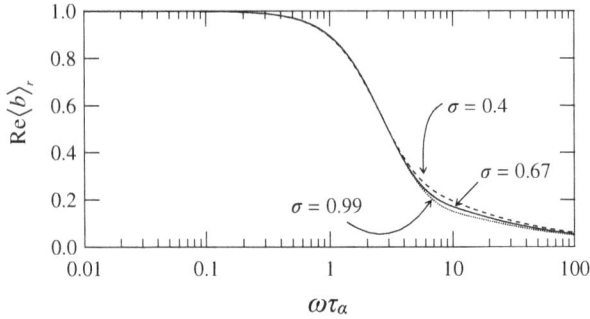

Figure 7.7: Relation between $\mathrm{Re}\langle b \rangle_r$ and $\omega\tau_\alpha$.

7.1.3 Experiments in standing wave field

The acoustic power amplification was also investigated in a standing wave field in an acoustic resonance tube [67], where the work source component W_{stand} due to the standing wave component of pressure oscillation dominates over W_{prog}. Figure 7.8 (a) illustrates the resonance tube used in the experiments. The tube was closed by a solid plate at one end, while it was connected to a loudspeaker at the other end. The length was 3.0 m and filled with air at ambient temperature and pressure. In the middle of the tube, the regenerator-heat exchanger assembly was installed. The driven frequency was set to the second harmonic frequency of the resonance tube, so that a pressure maximum was created at the position of the regenerator. A stack of metal screen meshes or a ceramic honeycomb catalyst support was used as a regenerator, whose length was 20 mm. The corresponding values of $\omega\tau_\alpha$ were 0.13 and 3.5. The temperature ratio between the hot and the cold ends of the regenerator was set to 1.9. The temperature ratio is given by $T_{out}/T_{in} = T_H/T_C$ or T_C/T_H, depending on the alignment of hot and cold heat exchangers.

Figures 7.8 (b) and (c) show the results of simultaneous measurements of pressure and velocity, where the pressure amplitude $|p_1|$ and cross-sectional mean velocity amplitude $|u_{1r}|$ are plotted as functions of the axial coordinate x in (b) and the phase lead ϕ of u_{1r} relative to p_1 is shown in (c). The phase ϕ was close to a traveling wave phasing at the position of the pressure maximum, but it changes steeply from

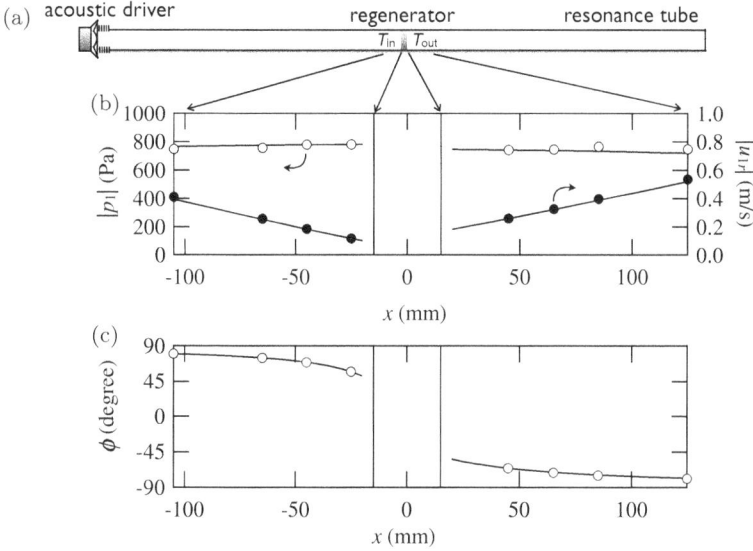

Figure 7.8: Experimental setup (a), and the acoustic field: (b) pressure amplitude $|p_1|$ and cross-sectional mean velocity amplitude $|u_{1r}|$, (c) phase ϕ of u_{1r} relative to p_1.

$\pi/2$ to $-\pi/2$ as x goes across it. In other words, in the standing-wave like field, the phase difference ϕ can be adjusted by the regenerator position. So this experiment made use of the regenerator position to investigate the ϕ dependence of G.

Figure 7.9 presents the acoustic power amplification gain $G = I_{out}/I_{in}$ as a function of the phase lead ϕ of the cross-sectional mean velocity to pressure. When the contribution from the work source component W_{stand} is considered, the gain G is expressed as

$$G = 1 + \frac{W_{prog} + W_{stand} + W_p + W_\nu}{I_C}\Delta x. \tag{7.20}$$

For the gas oscillating in the regenerator with $\omega\tau_\alpha = 0.13$, W_{stand} and W_p should be negligibly small because of the good thermal contact of the gas with the regenerator wall. Therefore, we can assume

$$G = 1 + \frac{W_{prog} + W_\nu}{I_C}\Delta x, \tag{7.21}$$

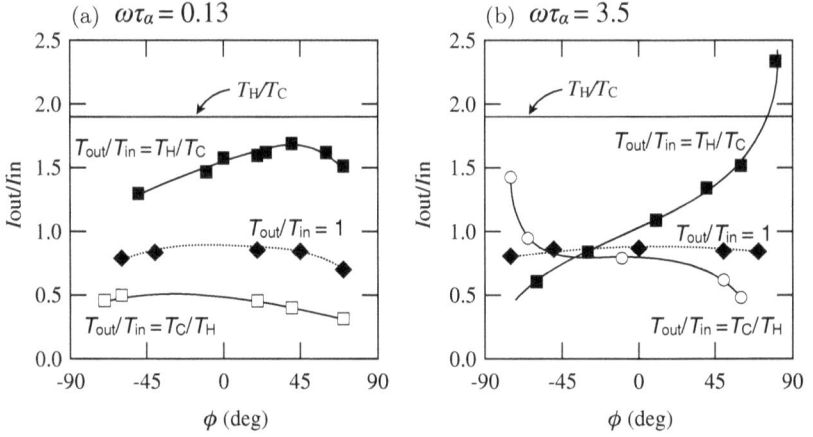

Figure 7.9: The acoustic power amplification gain G as a function of the phase lead ϕ of the cross-sectional mean velocity to pressure: (a) $\omega\tau_\alpha = 0.13$ and (b) $\omega\tau_\alpha = 3.5$.

when the temperature gradient is present, and

$$G = 1 + \frac{W_\nu}{I_C}\Delta x, \tag{7.22}$$

when it is absent. As shown in Fig. 7.9(a), G with $T_{out}/T_{in} = T_H/T_C$ goes above G with $T_{out}/T_{in} = 1$, reaching 89% of T_H/T_C of the regenerator ends at the highest. In Ceperley's theoretical analysis, G should be equal to the temperature ratio T_H/T_C. The results show that the viscous damping lowers G in reality.

In the case of the regenerator with $\omega\tau_\alpha = 3.5$, Fig. 7.9(b) shows that G rapidly changes as ϕ reaches $\pm\pi/2$ in both cases of $T_{out}/T_{in} = T_H/T_C$ and $T_{out}/T_{in} = T_C/T_H$. This characteristic behavior is attributable to the contribution of W_{stand}. As we learned in Chapter 6, W_{stand}/I reads

$$\frac{W_{stand}}{I} = -\beta\frac{dT_m}{dx}(\text{Im}\langle b\rangle_r)\tan\phi, \tag{7.23}$$

and hence, we see that $|G|$ goes to infinity as $\phi \to \pm\pi/2$. In the standing wave acoustic engine, I goes out of the stack from both sides, which means that ΔI is finite even when I does not enter the stack (see Chapter 4). This is in clear contrast to the traveling wave engine where I goes through the regenerator after being amplified by the temperature gradient. Thus, when W_{stand} governs the energy conversions, the gain G can be infinitely large, but when W_{prog} dominates, G is limited to T_H/T_C.

7.2 Temperature Gradient for Acoustic Power Production

Major function of a regenerator is promoting acoustic power production by enhancing the thermal interactions between oscillating fluid particles and solid walls. In this section, we discuss the condition required for the axial temperature gradient dT_m/dx to enable the acoustic power production. Specifically, we consider the case with a positive temperature gradient and investigate the minimum dT_m/dx leading to $w = 0$.

7.2.1 *Relation between the work source and the temperature gradient*

The work source w is decomposed into four components: W_{prog}, W_{stand}, W_p, and W_ν. Among them, W_p and W_ν are always negative. On the other hand, a sum of W_{prog} and W_{stand} can be positive. From the result of Chapter 6, $W_{prog} + W_{stand}$ is given by

$$W_{prog} + W_{stand} = \frac{1}{2}\beta\frac{dT_m}{dx}|p_1||u_{1r}|\{\text{Re}\langle b\rangle_r \cos\phi - \text{Im}\langle b\rangle_r \sin\phi\},$$

(7.24)

which is transformed by using $\Psi = \arg\langle b\rangle_r$ as

$$W_{prog} + W_{stand} = \frac{1}{2}\beta\frac{dT_m}{dx}|p_1||u_{1r}||\langle b\rangle_r| \cos(\phi + \Psi).$$

(7.25)

Therefore, $W_{prog} + W_{stand}$ is positive when $\cos(\phi + \Psi) > 0$ and $dT_m/dx > 0$. The subject of this section is to derive the threshold value of dT_m/dx that makes $w = 0$.

The work source is explicitly written as

$$w = W_{prog} + W_{stand} + W_p + W_\nu$$

$$= \frac{1}{2}\omega\rho_m \mathrm{Im}\left[\frac{1}{1-\chi_\nu}\right]|u_1 r|^2 + \frac{\omega}{2}(K_T - K_S)\chi_\alpha''|p_1|^2$$

$$+ \frac{1}{2}\beta\frac{dT_m}{dx}|p_1||u_{1r}||\langle b\rangle_r|\cos(\phi + \Psi). \tag{7.26}$$

Inserting thermodynamic relations

$$K_T - K_S = K_S(\gamma - 1) = \frac{\gamma - 1}{\rho_m c_S^2} \tag{7.27}$$

$$\beta = \frac{1}{T_m} \tag{7.28}$$

yields

$$w = \frac{\omega}{2c_S}|p_1||u_{1r}| \times \left\{\mathrm{Im}\left[\frac{1}{1-\chi_\nu}\right]\rho_m c_S\left|\frac{u_{1r}}{p_1}\right| + (\gamma - 1)\frac{\chi_\alpha''}{\rho_m c_S}\left|\frac{p_1}{u_{1r}}\right|\right.$$

$$\left. + \frac{c_S}{\omega T_m}\frac{dT_m}{dx}|\langle b\rangle_r|\cos(\phi + \Psi)\right\}. \tag{7.29}$$

For brevity, let us introduce two dimensionless quantities $\hat{z}$ and X:

$$\hat{z} = \frac{z}{\rho_m c_S} \tag{7.30}$$

$$X = \frac{x}{\bar{\lambda}}, \tag{7.31}$$

where $\hat{z}$ is the dimensionless acoustic impedance normalized by a characteristic impedance $\rho_m c_S$, whereas X denotes the dimensionless axial coordinate normalized by a characteristic length $\bar{\lambda} = c_S/\omega$. Therefore, we see

$$\hat{z} = \frac{1}{\rho_m c_S}\frac{p_1}{u_{1r}} \tag{7.32}$$

$$\frac{dT_m}{dX} = \bar{\lambda}\frac{dT_m}{dx}. \tag{7.33}$$

Use of $\hat{z}$ and X gives

$$
\begin{aligned}
w = \frac{\omega}{2c_S}|p_1||u_{1r}| \Bigg\{ &\mathrm{Im}\left[\frac{1}{1-\chi_\nu}\right]\frac{1}{|\hat{z}|} + (\gamma-1)\chi_\alpha''|\hat{z}| \\
&+ \frac{1}{T_m}\frac{dT_m}{dX}|\langle b\rangle_r|\cos(\phi+\Psi)\Bigg\}.
\end{aligned}
\tag{7.34}
$$

If we introduce a logarithmic temperature gradient

$$
\frac{d\log T_m}{dX} = \frac{1}{T_m}\frac{dT_m}{dX},
$$

the temperature gradient that makes $w=0$ is given by

$$
\frac{d\log T_m}{dX} = -\frac{\mathrm{Im}\left[\dfrac{1}{1-\chi_\nu}\right]\dfrac{1}{|\hat{z}|}}{|\langle b\rangle_r|\cos(\phi+\Psi)} - \frac{(\gamma-1)\chi_\alpha''|\hat{z}|}{|\langle b\rangle_r|\cos(\phi+\Psi)}.
\tag{7.35}
$$

In Eq. (7.35), we should note

$$
\cos(\phi+\Psi) > 0, \quad \mathrm{Im}\left[\frac{1}{1-\chi_\nu}\right] < 0, \quad \chi_\alpha'' < 0.
$$

Therefore, the first term and the second term on the right hand side are both positive, which means that the logarithmic temperature gradient $d\log T_m/dX$ should have a minimum when it is plotted as a function of $|\hat{z}|$ as shown in Fig. 7.10.

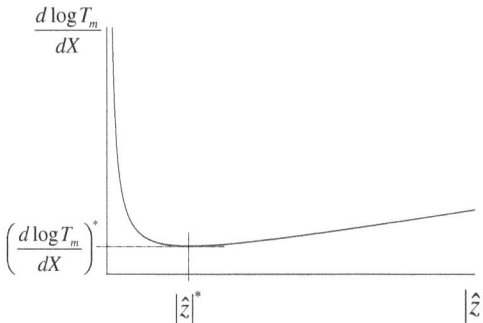

Figure 7.10: Relation between the logarithmic temperature gradient $d\log T_m/dX$ and the magnitude of dimensionless acoustic impedance $|\hat{z}|$.

The minimum is given by[1]

$$\left(\frac{d\log T_m}{dX}\right)^* = \frac{2\sqrt{(\gamma-1)\chi_\alpha'' \text{Im}\left[\frac{1}{1-\chi_\nu}\right]}}{|\langle b\rangle_r|\cos(\phi+\Psi)}, \qquad (7.36)$$

which is achieved with $\hat{z}^*$ given by

$$\hat{z}^* = \sqrt{\frac{\text{Im}\left[\frac{1}{1-\chi_\nu}\right]}{(\gamma-1)\chi_\alpha''}}\, e^{i\Psi}, \qquad (7.37)$$

where $\Psi = \arg\langle b\rangle_r$.

Figure 7.11 presents the minimum logarithmic temperature gradient $(d\log T_m/dX)^*$, and Figs. 7.12(a) and (b) show $|\hat{z}^*|$ (a) and $\phi^* = -\arg \hat{z}^*$. The minimum logarithmic temperature gradient $(d\log T_m/dX)^*$ is elevated with $\omega\tau_\alpha$ with a transitional region $1 < \omega\tau_\alpha < 10$. The magnitude $|\hat{z}^*|$ is always greater than unity, meaning that the specific acoustic impedance should be higher than the characteristic impedance $\rho_m c_S$. The phase ϕ^* shows a maximum value of $54°$ with $\omega\tau_\alpha \sim 7$ where $\sigma = 0.67$, and it approaches to

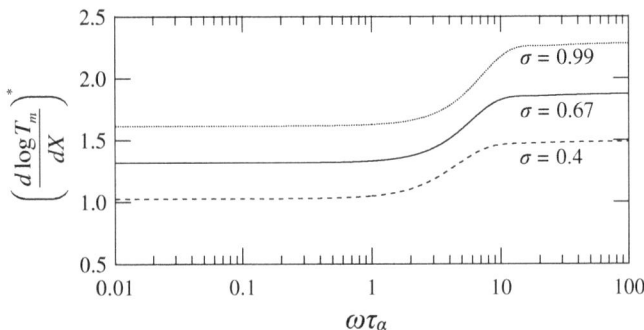

Figure 7.11: The minimum logarithmic temperature gradient $(d\log T_m/dX)^*$ as a function of $\omega\tau_\alpha$ ($\gamma = 1.66$).

---

[1]If the inequality of arithmetic and geometric means is used, the derivation would be easy.

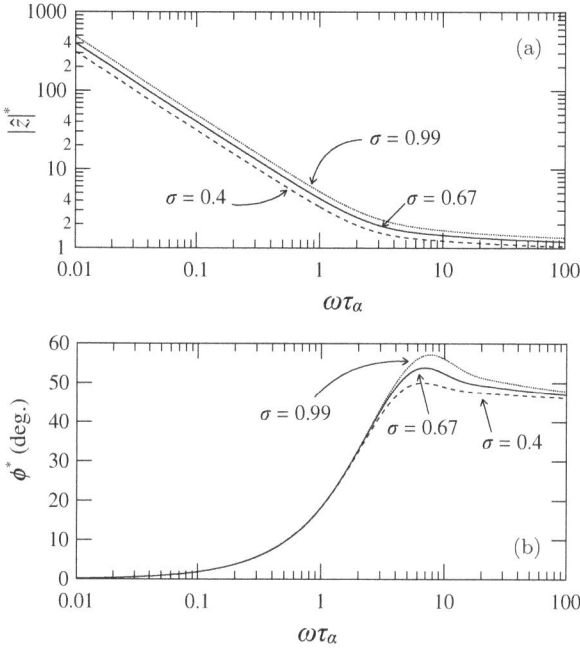

Figure 7.12: $|\hat{z}^*|$ (a) and $\phi^* = -\arg \hat{z}^*$ as a function of $\omega\tau_\alpha$ ($\gamma = 1.66$).

zero as $\omega\tau_\alpha$ is decreased. Thus, we are able to say that the phase of z should be tuned to enhance W_{prog} when $\omega\tau_\alpha$ is small, while it should be changed to include both the contribution from W_{prog} and W_{stand} when $\omega\tau_\alpha > 1$. Therefore, in order to induce thermoacoustic spontaneous gas oscillations with as small as possible temperature gradient, one should compare the specific acoustic impedance $\hat{z}$ that can be expected from the acoustic boundary conditions with $\hat{z}^*$ shown in Fig. 7.12.

7.2.2 Thermally induced spontaneous gas oscillations in resonance tube

We discuss here the relation between stack position in a resonance tube and the critical temperature gradient of spontaneous gas oscillations. In the standing wave field created in the resonance tube, the phase lead ϕ of u_{1r} relative to p_1 is close to $\pm\pi/2$, but we consider

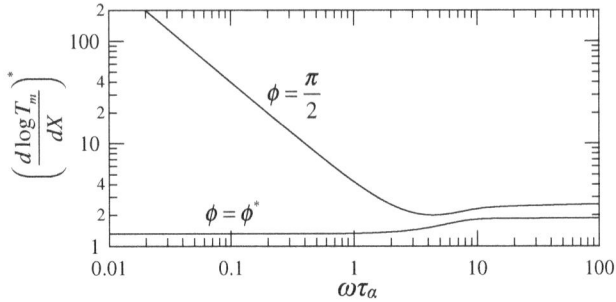

Figure 7.13: The minimum logarithmic temperature gradient $(d\log T_m/dX)^*$ as a function of $\omega\tau_\alpha$ ($\gamma = 1.66$ and $\sigma = 0.66$).

here the case with $\phi = \pi/2$, so that W_{stand} becomes positive when a positive temperature gradient is given to the stack. Inserting $\phi = \pi/2$ into Eq. (7.36) gives

$$\left(\frac{d\log T_m}{dX}\right)^* = -\frac{2\sqrt{(\gamma-1)\chi_\alpha''\mathrm{Im}\left[\dfrac{1}{1-\chi_\nu}\right]}}{|\langle b\rangle_r|\sin\Psi}. \tag{7.38}$$

Figure 7.13 compares the minimum logarithmic temperature gradient $(d\log T_m/dX)^*$ when the phase $\phi = \pi/2$ and $\phi = \phi^*(= -\Psi)$. The standing wave phase $\phi = \pi/2$ results in the higher $(d\log T_m/dX)^*$ than the optimum phase $\phi = \phi^*$, and takes a minimum value of 2.0 when $\omega\tau_\alpha = 4.4$. The magnitude of dimensionless acoustic impedance $|\hat{z}|^*$ is 2.2 with this value of $\omega\tau_\alpha$, as shown in Fig. 7.12(a). In other words, when the stack is located at the position that makes $|\hat{z}|^* = 2.2$, the temperature gradient necessary for the onset of gas oscillations is minimized.

In a resonance tube with length L and relatively large radius, the acoustic field can be safely assumed by

$$p_1 = -iz_S\frac{\cos[k_S(L-x)]}{\sin k_S L}U_0 \tag{7.39}$$

$$u_1 = \frac{\sin[k_S(L-x)]}{\sin k_S L}U_0 \tag{7.40}$$

as in Section 3.4. Letting $z = p_1/u_1 = 2.2\rho_m c_S$ yields the stack position, which turns out to be about $L/7$ or $L/8$ away from the hot end of the resonance tube. In the case of an open-closed resonance tube, this position roughly corresponds to half of the tube length. Let's recall these results when you plan to build a demonstration kit shown in Chapter 1.

The experiments on the resonance tube introduced in Section 4.6 offer another example to test the minimum logarithmic temperature gradient $(d \log T_m/dX)^*$ in Eq. (7.36). It is found that w changes from negative to positive between $T_H/T_C = 1.34$ and 1.51, where $T_C = 296$ K. From the stack length and oscillation frequency, the experimental logarithmic temperature gradient $(d \log T_m/dX)$ is estimated to be around 4.6. On the other hand, theoretical value would be estimated from Eq. (7.35) as

$$\frac{d \log T_m}{dX} = \frac{\mathrm{Im}\left[\dfrac{1}{1 - \chi_\nu}\right] + (\gamma - 1)\chi_\alpha''}{|\langle b \rangle_r| \sin \Psi}. \tag{7.41}$$

where we have used $\phi = \pi/2$ and $|\hat{z}| = 1$ because of the stack location. From experimental conditions, we have $\omega\tau_\alpha = 3.9$, $\sigma = 0.7$, $\gamma = 1.4$, and therefore, $d \log T_m/dX$ is calculated as 2.2, which is rather low compared with 4.6 derived earlier. The reason for the deviation may be due to the energy dissipations in hot and cold heat exchangers placed on the sides of the regenerator. It should also be noted that when one tries to estimate the temperature gradient to cause spontaneous gas oscillations, one should consider the energy dissipations in the resonance tube, as well as in heat exchangers. Therefore, a higher temperature gradient than Eq. (7.35) would be necessary in real systems. A more general treatment based on thermoacoustic theory is given by Ueda *et al.* [68]. According to their calculation results, the onset temperature gradient shows a U-shaped curve against $\omega\tau_\alpha$ with a minimum value at $\omega\tau_\alpha \sim 3$, which is in qualitative agreement with Fig. 7.13.

7.2.3   *Thermally induced spontaneous gas oscillations in looped tube*

The acoustic field built up in a looped tube thermoacoustic system is an interesting problem, because the periodic boundary condition allows a variety of acoustic fields. For a simple loop made of an empty tube without a regenerator and heat exchangers, the intrinsic oscillating modes include traveling waves going in a positive direction and in a negative direction. Let's examine the acoustic field in the looped tube thermoacoustic system by paying attention to how the system adjusts the combination of these traveling waves.

Ueda [68] investigated the acoustic field induced in a looped tube acoustic engine when the regenerators with different $\omega\tau_\alpha$ values were employed. The experimental setup is shown in Fig. 7.14(a). The tubes near the regenerator are made of glass tubes with numerous ducts in order to facilitate simultaneous measurement of pressure and velocity by a small pressure transducer and a laser Doppler velocimeter. Figure 7.14(b) presents the phase lead θ of pressure oscillations relative to displacement oscillations determined at the hot end of the regenerator. The phase θ changes from $65°$ with $\omega\tau_\alpha = 21$ to $85°$ with $\omega\tau_\alpha = 1.7$, and hence the phase ϕ of velocity relative to pressure changes from $25°$ to $5°$ as $\omega\tau_\alpha$ was reduced. This tendency coincides with ϕ^* shown in Fig. 7.12(b). Experimental results also indicate that the acoustic impedance increased with decreasing $\omega\tau_\alpha$, which is also in agreement with $|\hat{z}|^*$ in Fig. 7.12(a). Therefore, we see that the acoustic field is tuned so that the oscillations occur with as small as

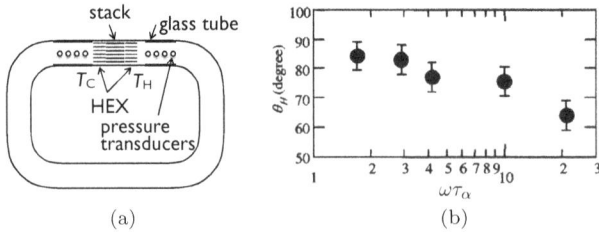

(a)   (b)

Figure 7.14: (a) Looped tube acoustic engine and (b) phase θ_H of displacement oscillations relative to pressure oscillations determined at the hot end of the regenerator [68].

possible a temperature gradient. It is interesting to note that such a self-tuning mechanism is naturally equipped with a looped tube acoustic engine. Taking advantage of that mechanism would be a key point for better functioning of the acoustic engine.

7.3 Thermal Efficiency of Energy Conversion in Regenerator

When a sufficiently high temperature gradient is provided to a regenerator, the work source w can become positive, and consequently the regenerator functions as a heat engine. The thermal efficiency of local energy conversion in the regenerator [121] is the main topic of this section. On the basis of results from thermoacoustic theory, we discuss the acoustic field that maximizes thermal efficiency. Also we describe a design method of the acoustic engines that aims at high thermal efficiency.

7.3.1 *Estimation of thermal efficiency*

We assume that a regenerator has a uniform cross-section for oscillating gas and has a positive temperature gradient dT_m/dx. The thermal efficiency η is given by the ratio of increase ΔI of the work flux density to the heat flux Q_H at the hot end of the regenerator as

$$\eta = -\frac{\Delta I}{Q_H}. \tag{7.42}$$

The negative sign is put because the sign of Q_H, flowing from hot to cold, is negative. If we use the work source w, ΔI is approximated by $w\Delta x$ where Δx is the regenerator length. Therefore,

$$\eta \sim -\frac{w\Delta x}{Q_H}. \tag{7.43}$$

The Carnot efficiency η_{Carnot} is given by using the hot end temperature T_H and the cold end temperature T_C as

$$\eta_{\text{Carnot}} = \frac{T_H - T_C}{T_H}, \tag{7.44}$$

which can then be assumed as

$$\eta_{\text{Carnot}} \sim \frac{dT_m}{dx} \frac{\Delta x}{T_H}. \tag{7.45}$$

The ratio ϵ of η to η_{Carnot} is given by

$$\epsilon = -\frac{T_H w}{Q_H \dfrac{dT_m}{dx}}. \tag{7.46}$$

In an ideal case where only the thermodynamically reversible processes occur, the entropy flux density $s = Q/T$ should be constant. Therefore, $T_H/Q_H = T_m/Q$ for the temperature T_m and the heat flux density Q at the middle of the flow channel segment Δx. In an actual case, s must increase its flow rate as it flows down. Therefore, $T_H/|Q_H| > T_m/|Q|$. Hence, if we substitute T_H/Q_H in Eq. (7.46) with T_m/Q, η will be underestimated, but provide a first approximation to the actual η. Then hereafter we consider the ratio of the local thermal efficiency relative to the Carnot efficiency:

$$\epsilon' = -\frac{w/Q}{\dfrac{d\log T_m}{dx}}. \tag{7.47}$$

The heat flux density Q is written as

$$Q = Q_{prog} + Q_{stand} + Q_D,$$

with

$$Q_{prog} = -\frac{1}{2}\beta T_m \text{Re}[g]|p_1||u_{1r}|\cos\phi,$$

$$Q_{stand} = -\frac{1}{2}\beta T_m \text{Im}[g]|p_1||u_{1r}|\sin\phi,$$

$$Q_D = \frac{1}{2\omega}\rho_m C_p \frac{dT_m}{dx} \text{Im}[g_D]\text{Re}\left[\frac{1}{1-\chi_\nu}\right]|u_{1r}|^2,$$

where ϕ represents the phase lead of u_{1r} relative to p_1. If the phase of g is represented by Θ, $\text{Re}[g] = |g|\cos\Theta$ and $\text{Im}[g] = |g|\sin\Theta$.

Therefore,

$$Q_{prog} + Q_{stand} = -\frac{1}{2}\beta T_m |p_1||u_{1r}||g|\cos(\phi - \Theta). \tag{7.48}$$

Then, we have

$$Q = -\frac{1}{2}\beta T_m |p_1||u_{1r}||g|\cos(\phi - \Theta) + \frac{1}{2\omega}\rho_m C_p \frac{dT_m}{dx}\mathrm{Im}[g_D]$$

$$\times \mathrm{Re}\left[\frac{1}{1 - \chi_\nu}\right]|u_{1r}|^2. \tag{7.49}$$

To further transform Q, we use thermodynamic relations for an ideal gas

$$\beta = \frac{1}{T_m}$$

$$\rho_m C_p = \frac{\rho_m c_S^2}{T_m(\gamma - 1)}$$

and also introduce dimensionless specific acoustic impedance $\hat{z}$ and dimensionless axial coordinate X

$$\hat{z} = \frac{z}{\rho_m c_S}$$

$$X = \frac{x}{\bar{\lambda}}$$

where $z = p_1/u_{1r}$ and $\bar{\lambda} = c_S/\omega$. The temperature gradient is rewritten as

$$\frac{dT_m}{dx} = \frac{1}{\bar{\lambda}}\frac{dT_m}{dX}.$$

The heat flux density Q is transformed as

$$Q = -\frac{1}{2}|p_1||u_{1r}|\left(|g|\cos(\phi - \Theta) - \frac{d\log T_m}{dX}\frac{\mathrm{Im}[g_D]}{\gamma - 1}\mathrm{Re}\left[\frac{1}{1 - \chi_\nu}\right]\frac{1}{|\hat{z}|}\right). \tag{7.50}$$

The work source w is expressed in Eq. (7.34) as

$$w = \frac{\omega}{2c_S}|p_1||u_{1r}| \left(\text{Im} \left[\frac{1}{1-\chi_\nu} \right] \frac{1}{|\hat{z}|} + (\gamma-1)\chi_\alpha''|\hat{z}| \right.$$
$$\left. + \frac{d\log T_m}{dX}|\langle b \rangle_r| \cos(\phi + \Psi) \right). \tag{7.51}$$

Inserting the expressions of Q and w into Eq. (7.47) gives the ratio of the local thermal efficiency relative to the Carnot efficiency, ϵ', as

$$\epsilon' = \frac{\text{Im} \left[\dfrac{1}{1-\chi_\nu} \right] \dfrac{1}{|\hat{z}|} + (\gamma-1)\chi_\alpha''|\hat{z}| + \dfrac{d\log T_m}{dX}|\langle b \rangle_r| \cos(\phi + \Psi)}{\dfrac{d\log T_m}{dX} \left(|g|\cos(\phi - \Theta) - \dfrac{d\log T_m}{dX} \dfrac{\text{Im}[g_D]}{\gamma-1}\text{Re}\left[\dfrac{1}{1-\chi_\nu} \right] \dfrac{1}{|\hat{z}|} \right).} \tag{7.52}$$

It is seen that ϵ' depends on the gas thermal properties of the Prandtl number σ and specific heat ratio γ, as well as $\omega\tau_\alpha$, logarithmic temperature gradient $d\log T_m/dX = \bar{\lambda}d\log T_m/dx$, and dimensionless specific acoustic impedance $\hat{z}$. Figure 7.15 (a)–(c) shows contour plots of ϵ' in the plane of $|\hat{z}|$ versus ϕ for $\omega\tau_\alpha$ values of 0.01, 0.05, 0.1, and 0.5, when $d\log T_m/dX = 20$.[2]

In the case of relatively small $\omega\tau_\alpha$ values of 0.01 and 0.05, ϵ' is symmetric with respect to the line $\phi = 0$. When $\omega\tau_\alpha = 0.1$, however, the maximum of ϵ' moves to the region with $\phi > 0$. This shift reflects the balance between W_{prog} and W_{stand}; the contribution of W_{stand} becomes large near $\omega\tau_\alpha \sim \pi$, whereas it is negligibly small when $\omega\tau_\alpha \ll \pi$. We see a region with $\epsilon' > 0.8$, when $|\hat{z}|$ is high and ϕ is close to zero, and when $\omega\tau_\alpha$ is as small as 0.05. Therefore, it is critically important to map out ϵ' when we design a traveling-wave thermoacoustic prime mover. Also we should note that too high $|\hat{z}|$ results in the decrease of ϵ' when $\omega\tau_\alpha$ is 0.5. Therefore, the location of a stack must be chosen carefully in a standing-wave thermoacoustic

---

[2]Choice of the logarithmic temperature gradient needs some consideration, because w can be negative for small dT_m/dx.

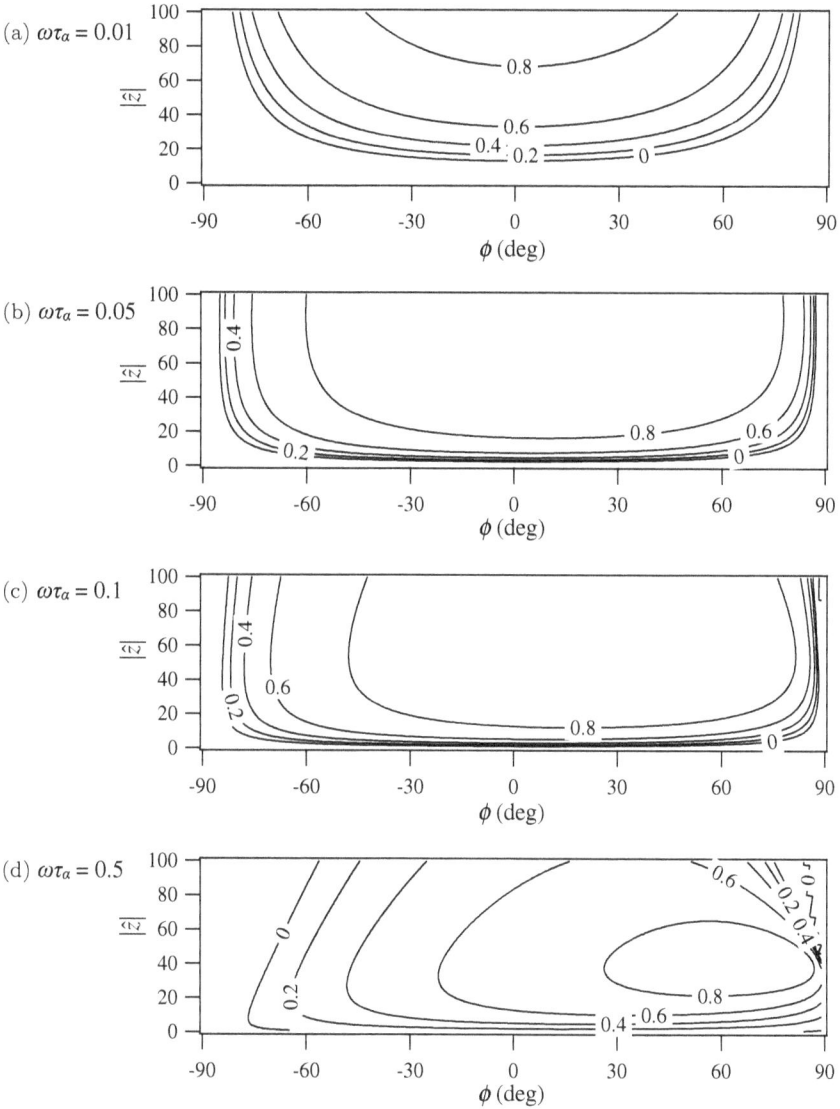

Figure 7.15: Contour plot of ϵ', when $\gamma = 5/3$, $\sigma = 0.66$ and $d\log T_m/dX = 20$; $\hat{z}$ is the dimensionless specific acoustic impedance, whereas ϕ denotes phase of u_{1r} relative to p_1.

prime mover, so that it is positioned neither too close to the pressure
maximum, nor too close to the pressure minimum.

7.3.2 *Looped tube engine with branch resonator*

In Stirling engines, solid pistons and displacers are mechanically
tuned in such a way that the desired acoustic fields are established.
In acoustic engines, the acoustic fields are tuned by appropriately
designing the tube configuration. Figure 7.16(a) presents a looped
tube traveling wave engine with a branch tube developed by Backhaus and Swift in Los Alamos Laboratory [28]. They succeeded in
extracting the acoustic power of 700 W from the engine through
the branch tube. The thermal efficiency reached 30%, when it was
deduced from the extracted acoustic power and the heat supplied
to the engine. This efficiency corresponds to 42% of Carnot efficiency, and it is as high as internal combustion engines. Tijani in
Energy research Center of the Netherlands (ECN) also developed
the looped tube engine with branch tube and achieved thermal
efficiency reaching 48% of Carnot efficiency [29]. The reason for
these high efficiencies can be found in the acoustic field that they
created.

Figure 7.16(b) depicts the electrical circuit analogy of the
looped tube of their engines, where L and C denote the equivalent

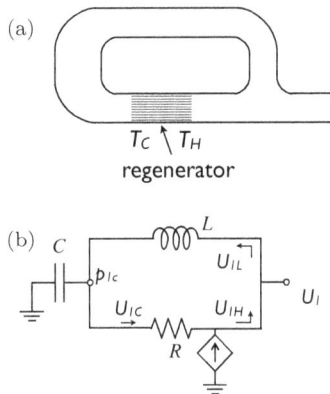

Figure 7.16: Schematic of looped tube traveling wave engine with (a) a branch
tube and (b) its electrical circuit analogy.

inductance and capacitance of the working gas. The differentially heated regenerator is represented by a resistance and a parallel current source. According to Ceperley's discussion, the volume velocity is amplified by a temperature gradient along the regenerator. Therefore, the volume velocities U_{1H} and U_{1C} at hot and cold ends of the regenerator are linked by $U_{1H}/U_{1C} = T_H/T_C$. This means that regenerator is equipped with a volume velocity source which provides $(T_H/T_C - 1)U_{1C}$. If the volume velocity passing though the inductance is denoted by U_{1L} and the pressure at the cold end of the regenerator is denoted by p_{1C}, we have two circuit equations

$$RU_{1C} = -i\omega LU_{1L} \qquad (7.53)$$

$$U_{1L} - U_{1C} = i\omega C p_{1C}. \qquad (7.54)$$

By eliminating U_{1L} from Eqs. (7.53) and (7.54), we obtain the acoustic impedance $Z_C = p_{1C}/U_{1C}$ at the cold end of the regenerator as

$$Z_C = \frac{R}{\omega^2 LC}\left(1 + i\frac{\omega L}{R}\right). \qquad (7.55)$$

This result indicates that a traveling wave phase and a high acoustic impedance, essential for a high thermal efficiency, are achieved if $R \gg \omega L$. As we see from Fig. 7.15, rather high thermal efficiency is expected with $\phi = -45°$ if $\omega\tau_\alpha$ is as low as 0.01. In this case, the condition is modified to $R \sim \omega L$. If we set $R = \omega L$, the magnitude of acoustic impedance at the cold end is given by $|Z_C| = \sqrt{2}/(\omega C)$. Since the capacitance is given as $C = V/(\gamma p_m)$ by using the cavity volume V, we are able to estimate the ratio of $|Z_C|$ over the characteristic impedance $\rho_m c_S$. Here, ω would be predicted from the natural frequency of the engine determined by the size of looped tube and branch tube, and type of working gas. It seems that Backhaus made such a tabletop calculation before actually building the experimental setup. The validity was indeed verified by the thermal efficiency achieved by their engine. A detailed measurement result on the acoustic field of the looped tube engine is given in references [69, 70].

7.4 Appendix

7.4.1 *How to build a loaded looped tube engine*

The looped tube engine in Fig. 7.16 can serve as an electric generator or a heat driven cooler, if an electroacoustic convertor like a linear alternator [71] or a looped tube acoustic cooler is connected to the end of the resonance tube [72]. In order to make the total size of the system compact, the branch tube length should be reduced as much as possible, but such modification may result in extremely high temperature difference for its operation. Hatori discussed how to choose acoustic load when one tries to connect a looped tube engine and a load via branch tube [73].

Hatori separated the loaded engine at the middle of the branch tube into two parts: engine part and load part. He considered that the pressure and volume velocity at the junction of these two parts must be equal to each other, otherwise the combined engine would not start oscillating because of the discontinuity. This condition means that these two parts should possess the same acoustic impedance at a certain frequency, if they could be treated as linear systems. In other words, the combination of frequency and temperature difference that makes the impedances coincide with each other gives the operating conditions of the system made of two subsystems.

In order to verify his idea, he made measurements of acoustic impedances of two parts. An acoustic driver was connected to each part to excite periodically steady oscillation of the working gas. The acoustic pressure and velocity were then measured at the junction positions, while varying the driving frequency. For the engine part, the temperature difference along the regenerator was also taken as the experimental parameter.

Figure 7.17 presents the measured specific acoustic impedance in three-dimensional space composed of complex plane and frequency axis. The plane represents the acoustic impedance of the engine part, while the curves represent the impedance of two types of loads (load A and load B). It is seen that the plane and the curve intersect at a point. The frequency and temperature difference of that crossing point give the operating condition of the combined system, which

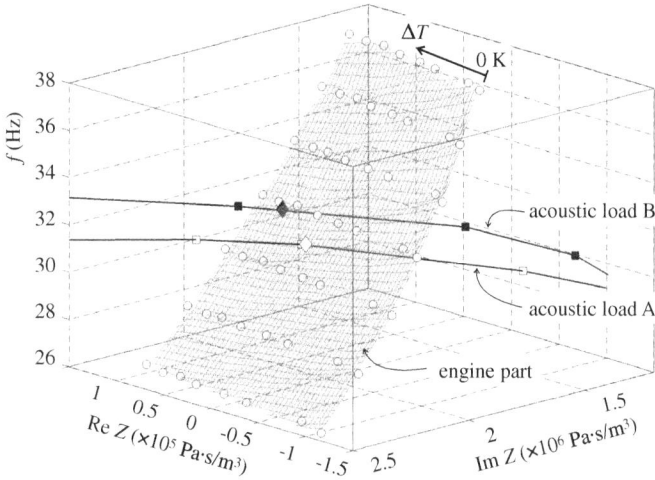

Figure 7.17: Complex specific acoustic impedances of the engine part and the load part. The engine impedance is shown by a plane, while the load impedance is drawn by curves.

was later confirmed by experiments done on the combined engine. Therefore, we are able to predict the operating conditions before we actually build the combined engine. Reference [74] documents the analysis of oscillation amplitude based on the acoustic impedance of subsystems.

7.4.2 *Liquid-piston looped tube engine*

One of the Stirling engines that use oscillating liquid columns is Fluidyne introduced in Chapter 1. Here you see another example of that type of Stirling engine [75, 76] in Fig. 7.18. This engine is made of three U-tubes of liquid column connected together by three gas columns containing regenerator and heat exchangers.

Recently, the analysis of liquid piston looped tube engine was made by modeling the engine with masses and springs; the masses stood for liquid column, and the springs expressed the effect of gas column (gas spring) and gravitational acceleration. Natural oscillation mode of the engine presented from the solution of equations of motion have shown that traveling wave phasing is guaranteed

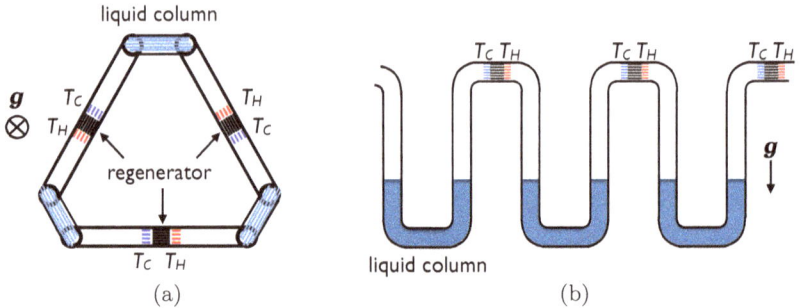

Figure 7.18: Schematic of liquid-piston looped tube engine.

if the engine is symmetrically made of the same U-tubes [77]. Also, the magnitude of acoustic impedance was shown to be easily controlled by the level of liquids that form U-tubes. The experimental observations have confirmed the validity of this theoretical analysis [78]. This result suggests that the acoustic field in the gas column is hardly influenced by the resistance of the regenerator. Thus, the design of the engine would be easier than the gas-based acoustic engine where the regenerator type and location impact the acoustic field.

The further test of this type of liquid piston engine is on-going. Comparison with Fluidyne is interesting in terms of output power and thermal efficiency, because they are both no-moving part Stirling engine.

7.5 Problems

1 Consider an equivalent mass-spring system of the liquid piston Stirling engine as shown in Fig. 7.19. The model is made of three unit sections each consisting of a point mass and two springs. The point mass with mass m represents the liquid column. The leaf spring with spring constant k_g provides the restoring force due to the gravitational force g, whereas the coil spring with k_a denotes that due to the bulk modulus of the gas. In this mass-spring model, the regenerators and heat exchangers are neglected.

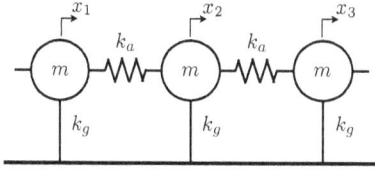

Figure 7.19: Schematic of liquid-piston looped tube engine.

The displacement from the equilibrium position is denoted by x_j $(j = 1, 2, 3)$.

(1) Derive the equations of motion of the system.
(2) Determine the natural frequency of the system.
(3) Obtain the specific acoustic impedance in the middle of the adjacent liquid columns.

Answers

1 (1) The equations of motion of the system are given by

$$m\frac{d^2 x_1}{dt^2} = k_a x_3 - (2k_a + k_g)x_1 + k_a x_2,$$

$$m\frac{d^2 x_2}{dt^2} = k_a x_1 - (2k_a + k_g)x_2 + k_a x_3,$$

$$m\frac{d^2 x_3}{dt^2} = k_a x_2 - (2k_a + k_g)x_3 + k_a x_1.$$

(2) By assuming $x_j = \mathrm{Re}(X_j e^{i\Omega t})$, and inserting into the equations of motion yields

$$-m\Omega^2 X_1 = k_a X_3 - (2k_a + k_g)X_1 + k_a X_2,$$
$$-m\Omega^2 X_2 = k_a X_1 - (2k_a + k_g)X_2 + k_a X_3,$$
$$-m\Omega^2 X_3 = k_a X_2 - (2k_a + k_g)X_3 + k_a X_1.$$

From the periodic boundary condition, X_j must be a periodic function $(j = 1, 2, 3)$;

$$X_j = a_q \exp\left(i\frac{2\pi q}{n}j\right), \tag{7.56}$$

where q is an integer and a_q is a constant. By substituting into the equations of motion, we obtain the frequency equation of the system:

$$\Omega_q^2 = \frac{k_g}{m} + \frac{2k_a}{m}\left(1 - \cos\frac{2\pi q}{n}\right).$$

The number of independent oscillation modes is 3. Hence the subscript q takes the values of 0, 1, 2.

(3) The velocity u' of the gas particle in the central part of the gas column would be approximated by the average velocity of the adjacent liquid columns. Therefore,

$$u' = \frac{i\Omega_q}{2}(X_j + X_{j+1}).$$

The pressure is expressed by

$$p' = \frac{k_a}{A}(X_j - X_{j+1}).$$

By inserting the solution of X_j, we obtain

$$u' = \frac{i\Omega_q a_q}{2}e^{i\frac{2\pi q j}{n}}\left(1 + e^{i\frac{2\pi q}{n}}\right),$$

$$p' = \frac{k_a a_q}{A}e^{i\frac{2\pi q j}{n}}\left(1 - e^{i\frac{2\pi q}{n}}\right).$$

The specific acoustic impedance is given by

$$z = \frac{2k_a}{iA\Omega_q}\frac{1 - e^{i\frac{2\pi q}{n}}}{1 + e^{i\frac{2\pi q}{n}}} = -\frac{2k_a}{A\Omega_q}\tan\left(\frac{\pi q}{n}\right). \qquad (7.57)$$

The specific acoustic impedance z is a real number, which means that the pressure p and velocity u oscillate temporally in phase. In other words, the traveling wave phasing is always achieved in the looped-tube liquid-piston engine.

Chapter 8

Heat Flow

Heat flux density caused by oscillating fluid parcels has three components: Q_{prog} due to traveling wave component of pressure oscillation, Q_{stand} due to standing wave component of pressure oscillation, and Q_D due to displacement oscillation in a longitudinal temperature gradient. Heat flow components Q_{prog} and Q_{stand} contribute to cooling power in coolers, whereas Q_D enhances heat transfer from hot to cold in dream pipes. In this chapter, cooler performance is discussed in terms of the heat flow components.

8.1 Acoustic Cooler

Consider oscillating gas parcels in a regenerator installed in a tube shown in Fig. 8.1. The regenerator is sandwiched by hot and cold heat exchangers. The tube is thermally insulated from outside except for the heat exchangers. Also we assume that the tube is so wide that the gas oscillates adiabatically, while the regenerator possesses narrow channel capable of maintaining thermal contact of the gas with channels wall. By exciting longitudinal acoustic gas oscillations using an acoustic driver like a loudspeaker or an acoustic engine, the heat flow $\widetilde{Q}$ is induced in the regenerator. The heat exchangers on the sides of the regenerator serve as the inlet and outlet of $\widetilde{Q}$. Thus, the heat is pumped up from heat source to heat sink through the regenerator and heat exchangers. The cooling power of an acoustic cooler refers to the heat absorbed by the cold heat exchanger, which can be approximated by the heat flow at the cold

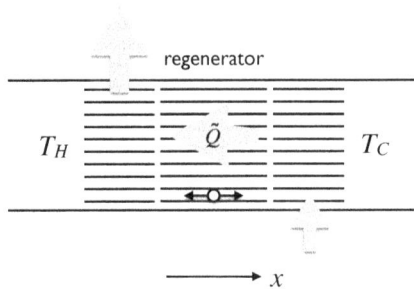

Figure 8.1: Schematic illustration of a regenerator sandwiched by hot and cold heat exchangers. The tube is thermally insulated from outside except for heat exchangers.

end of the regenerator, while the heating power of an acoustic heat pump refers to heat released from the hot heat exchanger, which can be substituted by heat flow at the hot end of the regenerator.

When an axial coordinate x is directed from the hot end (temperature T_H) to the cold end (temperature T_C), heat flow $\widetilde{Q}$ should flow in a negative direction ($\widetilde{Q} < 0$) when the regenerator serves as a heat pump or a cooler. The heat flow $\widetilde{Q}$ is expressed by a sum of heat flux density Q due to oscillating gas and Q_κ due to conduction heat by thermal conductivity of the gas and the solid parts

$$\widetilde{Q} = AQ + A_\kappa Q_\kappa \qquad (8.1)$$

where A denotes the gas-filled cross-sectional area, A_κ represents the cross-sectional area responsible for conduction heat, and

$$Q_\kappa = -\kappa \frac{dT_m}{dx}. \qquad (8.2)$$

If the solid components have negligibly small cross-sectional area, we can assume $A \approx A_\kappa$ and the thermal conductivity κ of the gas should be considered. As discussed in Chapter 6, heat flux density Q is expressed by a sum of Q_{prog} due to traveling wave component of pressure oscillation, Q_{stand} due to standing wave component of pressure oscillation, and Q_D due to displacement oscillation in a longitudinal temperature gradient. They are given for viscous fluid as follows.

$$Q_{prog} = -\frac{1}{2}\beta T_m \text{Re}[g]|p_1||u_{1r}|\cos\phi \tag{8.3}$$

$$Q_{stand} = -\frac{1}{2}\beta T_m \text{Im}[g]|p_1||u_{1r}|\sin\phi \tag{8.4}$$

$$Q_D = \frac{1}{2\omega}\rho_m C_p \frac{dT_m}{dx}\text{Im}[g_D]\text{Re}\left[\frac{1}{1-\chi_\nu}\right]|u_{1r}|^2, \tag{8.5}$$

with

$$g = \frac{\chi_\alpha - \chi_\nu^\dagger}{(1+\sigma)(1-\chi_\nu^\dagger)} = \frac{(1-\chi_\nu)(\chi_\alpha - \chi_\nu^\dagger)}{(1+\sigma)|1-\chi_\nu|^2} \tag{8.6}$$

and

$$g_D = \frac{\chi_\alpha - \chi_\nu^\dagger - (1+\sigma)\chi_\nu + (1+\sigma)\text{Re}[\chi_\nu]}{(1-\text{Re}[\chi_\nu])(1-\sigma^2)}. \tag{8.7}$$

If we use the cross-sectional mean displacement amplitude ξ_{1r} in place of u_{1r}, Q_{prog}, Q_{stand}, and Q_D are expressed by

$$Q_{prog} = -\frac{\omega}{2}\beta T_m \text{Re}[g]|p_1||\xi_{1r}|\sin\theta \tag{8.8}$$

$$Q_{stand} = -\frac{\omega}{2}\beta T_m \text{Im}[g]|p_1||\xi_{1r}|\cos\theta \tag{8.9}$$

$$Q_D = \frac{\omega}{2}\rho_m C_p \frac{dT_m}{dx}\text{Im}[g_D]\text{Re}\left[\frac{1}{1-\chi_\nu}\right]|\xi_{1r}|^2. \tag{8.10}$$

In the following section, we discuss the cooling performance of coolers introduced in Chapter 1 on the basis of the oscillation induced heat flow

$$Q = Q_{prog} + Q_{stand} + Q_D.$$

8.1.1 *Resonance tube cooler*

Merkli and Thomann [13] made some experiments of forced gas oscillations in a resonance tube. They put an acoustic driver to the

end of a gas filled resonance tube and closed the other end by a
solid plate. The driving frequency was tuned to a resonance frequency
of the gas column. They inserted the resonance tube into the larger
diameter tube in order to attain a good thermal insulation from the
outside, and measured the temperature along the resonance tube
wall. Just after the start of forced gas oscillations, reduction of
temperature was observed near the velocity maximum in the middle
of the resonator, whereas the increase was observed near the ends.
This was the first demonstration of cooling by sound waves. Let's
investigate the mechanism of the sound cooling.

For brevity, we assume a lossless acoustic field (see Fig. 8.2)
as we have discussed in Chapter 3, where the phase lead ϕ of the
velocity relative to pressure switches between $-90°$ and $90°$ at the
velocity (pressure) maximum and minimum. Owing to the phase
change, Q_{stand} changes its flow direction; Q_{stand} flows in the positive
x direction with $\phi = 90°$, whereas it flows in the negative x direction
with $\phi = -90°$. The contribution from Q_{prog} should be negligibly
small as the actual acoustic field is approximated by a standing
wave field. Thus the observed temperature distribution is interpreted
as a result of Q_{stand}. They also documented that after the initial
buildup of the temperature gradient, the system reached a more
uniform temperature state elevated from room temperature. The
energy dissipation due to viscosity and thermal conductivity of the
gas, and the influence of Q_D and Q_κ, both transporting heat from
hot to cold, would have masked the result of heat pumping by Q_{stand}.

A key component to enhance Q_{stand} was demonstrated by
Wheatley et al. [79]. They made a *thermoacoustic couple*, which is
a stack of parallel plates much shorter than the sound wavelength
and with relatively small spacing comparable to thermal boundary
layer thickness of the gas, and inserted it into the resonance tube. The
thermoacoustic couple is also referred to as *stack*. The stack promotes
thermal interactions between the oscillating gas and the solid plates
resulting in the increase of Q_{stand}. In the periodically steady state,
Q_{stand} should be balanced with Q_D and Q_κ; $A(Q_{stand} + Q_D) + A_\kappa Q_\kappa = 0$. According to their analysis, this relation is approximated
by $AQ_{stand} + A_\kappa Q_\kappa = 0$. By inserting Eqs. (8.2) and (8.4) into

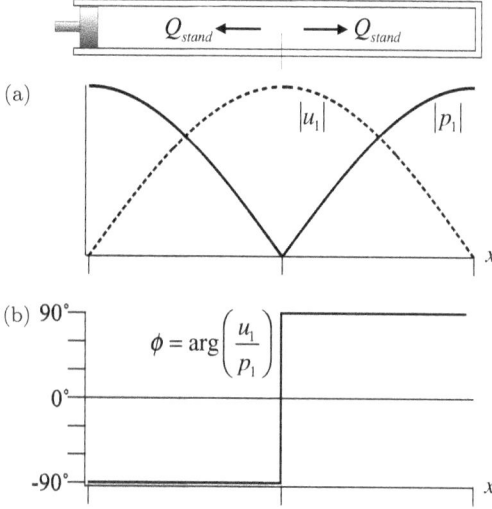

Figure 8.2: Schematic illustration of a resonance tube and the acoustic field: (a) velocity amplitude $|u_1|$ and pressure amplitude $|p_1|$, (b) phase lead ϕ of velocity relative to pressure.

$AQ_{stand} + A_\kappa Q_\kappa = 0$, the temperature gradient of the stack is given by

$$\frac{dT_m}{dx} = -\frac{A\beta T_m \text{Im}[g]}{2\kappa A_\kappa}|p_1||u_{1r}|\sin\phi. \tag{8.11}$$

The pressure and cross-sectional mean velocity are expressed by

$$p_1 = -iz_S\frac{\cos[k_S(L-x)]}{\sin k_S L}U_0 \tag{8.12}$$

$$u_{1r} = \frac{\sin[k_S(L-x)]}{\sin k_S L}U_0, \tag{8.13}$$

and hence we have

$$\frac{dT_m}{dx} = -\frac{A\beta T_m z_S \text{Im}[g]}{4\kappa A_\kappa}\frac{\sin[2k_S(L-x)]}{\sin^2 k_S L}U_0^2. \tag{8.14}$$

This result means that the temperature gradient becomes the greatest when the stack is located in the middle ($x = L/2$) of the pressure maximum and velocity minimum, whereas it approaches zero when the stack goes in the vicinity of the pressure maximum and

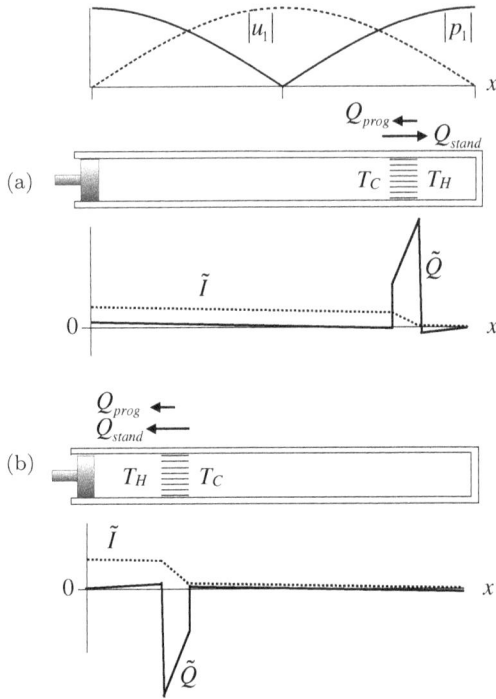

Figure 8.3: Energy flow diagram of resonance tube cooler. The heat flow component Q_{stand} is directed to pressure maximum, whereas Q_{prog} goes in the opposite direction to $\tilde{I}$.

minimum. Also, it should be noted that the higher temperature side of the stack faces the pressure maximum, as explained in Chapter 6. If the contribution from Q_D becomes unavoidable, the stack location should be shifted towards the velocity minimum to enlarge the temperature gradient.

The energy flow diagram of the resonance tube cooler is shown in Fig. 8.3, where a half wavelength standing wave is assumed. In Fig. 8.3(a), the stack is located close to the closed end. In this configuration, Q_{stand} and Q_{prog} go to opposite directions, so that they cancel each other. On the other hand, in the configuration in (b) where the stack is placed near the acoustic driver, Q_{stand} and Q_{prog} go in the same direction towards the driver. Therefore, the higher cooling performance is expected in (b) than (a).

The Los Alamos group was encouraged by their successful results, and challenged to build a heat-driven cooler called *beer cooler* [80]. Garrett in Pennsylvania State University developed a standing wave cooler, which was flown on the Space Shuttle Discovery [81].

8.1.2 *Looped tube acoustic cooler*

Constructed using a looped tube is a traveling wave component based acoustic cooler, where Q_{prog} plays a central role in cooling performance. As we have seen in Chapter 4, the work flow can be enlarged by replacing the resonance tube with a looped tube, because of the traveling wave going around the loop. Since the flow rate of Q_{prog} is proportional to I, the cooling power should also be enlarged in the looped tube acoustic cooler.

Swift, one of successors of Wheatley, developed a looped tube cooler and reported its performance [28]. If we express the work flow in the regenerator placed in the loop by I, Q_{prog} would be given by

$$Q_{prog} = -\frac{1}{2}\beta T_m \text{Re}[g]|p_1||u_{1r}|\cos\phi.$$

The heat flux density Q_D is the main heat loss

$$Q_D = \frac{1}{2\omega}\rho_m C_p \frac{dT_m}{dx}\text{Im}[g_D]\text{Re}\left[\frac{1}{1-\chi_\nu}\right]|u_{1r}|^2.$$

Letting $Q_{prog} + Q_D = 0$ gives

$$\frac{dT_m}{dx} = \omega\frac{\beta T_m}{\rho_m C_p}\frac{\text{Re}[g]}{\text{Im}[g_D]\text{Re}\left[\dfrac{1}{1-\chi_\nu}\right]}\text{Re}[z],$$

which indicates that placing the regenerator near the velocity minimum results in greater temperature gradient. However, the velocity minimum is not fixed in space but shifts with the regenerator position. Ueda et al. [82] investigated the acoustic field in the looped tube cooler when the regenerator length, pore size, and location are changed. Furthermore, they presented a possibility of achieving 60% of the Carnot COP in the optimally designed looped tube acoustic cooler.

Various types of acoustic coolers have been proposed. In a heat driven cooler of looped tube type, Yazaki [39] placed the secondary regenerator with $\omega\tau_\alpha \sim 3$ at a position a bit away from the velocity minimum, to make Q_{stand}, as well as Q_{prog} contribute to lowering temperatures. Garrett in Pennsylvania State University [83] and Tijani in Energy research Center of the Netherlands (ECN) [35] developed coaxial-tube acoustic cooler. In this type of cooler, I circulates through the inner tube and the annular region between the outer and inner tube. Hasegawa et al. built a heat driven cooler by connecting a looped tube engine and a looped tube cooler with a branch resonator [84]. By installing three regenerators in the engine loop, they achieved a cooling temperature of $-100°C$ with hot end temperature of $300°C$. These acoustic coolers are anticipated as a novel cooling technology without Freon gases.

8.2 COP of Regenerator

This section discusses COP of energy conversion in the regenerator when it serves as a heat pump. We consider a regenerator of length Δx and uniform gas-occupied cross-section, and assume a negative temperature gradient dT_m/dx along it. When Δx is much shorter than the wavelength of sound waves, decrease ΔI of work flux density I between the ends of the regenerator is approximated by $\Delta I \approx w\Delta x$, where w denotes the work source. In the case of an acoustic cooler, the coefficient of performance COP_C is expressed by using the heat flux density Q_C at the cold end as

$$\mathrm{COP}_C = \frac{Q_C}{\Delta I} \approx \frac{Q_C}{w\Delta x}. \tag{8.15}$$

The Carnot COP being a thermodynamic upper limit of COP, is expressed by

$$\mathrm{COP}_{C,Carnot} = \frac{T_C}{T_H - T_C} \approx \frac{T_C}{-(dT_m/dx)\Delta x}. \tag{8.16}$$

Let's consider the ratio ϵ_C

$$\epsilon_C = \frac{\text{COP}_C}{\text{COP}_{C,Carnot}} = -\frac{Q_C}{wT_C}\frac{dT_m}{dx} \tag{8.17}$$

as a measure of the efficiency of energy conversion.

In an ideal regenerator, the entropy flux density is kept constant, because of the absence of irreversible thermodynamic processes. Therefore, Q_C/T_C should be equal to Q/T_m at the middle of the regenerator: $Q_C/T_C = Q/T_m$. In an actual regenerator, the entropy flux density should increase its flow rate as it flows through the regenerator. Hence, $|Q_C|/T_C < |Q|/T_m$. As a result, substituting Q_C/T_C with Q/T_m overestimates ϵ_C, but it would give an approximate value as long as we think of a regenerator with relatively high COP_C. Then we consider the ratio of the local COP_C relative to $\text{COP}_{C,Carnot}$:

$$\epsilon'_C = -\frac{Q}{w}\frac{d\log T_m}{dx}. \tag{8.18}$$

The heat flux density Q can be written as

$$Q = -\frac{1}{2}|p_1||u_{1r}|\left(|g|\cos(\phi - \Theta) - \frac{d\log T_m}{dX}\frac{\text{Im}[g_D]}{\gamma - 1}\text{Re}\left[\frac{1}{1 - \chi_\nu}\right]\frac{1}{|\hat{z}|}\right), \tag{8.19}$$

with

$$\Theta = \arg g \tag{8.20}$$

as we have seen in Chapter 7.

The work source w is expressed as

$$w = \frac{\omega}{2c_S}|p_1||u_{1r}|\left(\text{Im}\left[\frac{1}{1 - \chi_\nu}\right]\frac{1}{|\hat{z}|} + (\gamma - 1)\chi''_\alpha|\hat{z}|\right.$$
$$\left. + \frac{d\log T_m}{dX}|\langle b\rangle_r|\cos(\phi + \Psi)\right) \tag{8.21}$$

with

$$\Psi = \arg\langle b\rangle_r. \tag{8.22}$$

Thus, ϵ'_C is explicitly written as

$$\epsilon'_C = \frac{d\log T_m}{dx} \frac{|g|\cos(\phi - \Theta) - \frac{d\log T_m}{dX}\frac{\mathrm{Im}[g_D]}{\gamma - 1}\mathrm{Re}\left[\frac{1}{1 - \chi_\nu}\right]\frac{1}{|\hat{z}|}}{\mathrm{Im}\left[\frac{1}{1 - \chi_\nu}\right]\frac{1}{|\hat{z}|} + (\gamma - 1)\chi''_\alpha|\hat{z}| + \frac{d\log T_m}{dX}|\langle b\rangle_r|\cos(\phi + \Psi)}. \tag{8.23}$$

It is seen that ϵ'_C depends on the gas thermal properties of Prandtl number σ and specific heat ratio γ, as well as $\omega\tau_\alpha$, logarithmic temperature gradient $d\log T_m/dX = \bar{\lambda}d\log T_m/dx$, and dimensionless specific acoustic impedance $\hat{z}$. Figures 8.4 (a)–(d) show contour plots of ϵ'_C in the plane of $|\hat{z}|$ versus ϕ for $\omega\tau_\alpha$ values of 0.01, 0.05, 0.1, and 0.5, when $\gamma = 5/3$, $\sigma = 0.66$, and $d\log T_m/dX = -20$.

In the case of relatively small $\omega\tau_\alpha$ values of 0.01 and 0.05, ϵ'_C is symmetric with respect to the line $\phi = 0$. When $\omega\tau_\alpha = 0.1$, however, the maximum of ϵ'_C moves to the region with $\phi < 0$. This shift comes from the fact the Q_{prog} and Q_{stand} flow in the same direction when $\phi < 0$. We see a region with $\epsilon'_C > 0.8$, when $|\hat{z}|$ is high and ϕ is close to zero, and when $\omega\tau/\alpha$ is as small as 0.05. Also we should note that too high $|\hat{z}|$ results in the decrease of ϵ'_C when $\omega\tau_\alpha$ is 0.5. Therefore, it would be important to map out ϵ'_C when we design an acoustic cooler.

8.3   Cooling Performance of GM Refrigerator

The thermoacoustic theory was originally aimed at accounting for thermally induced gas oscillations, but it was first applied to understanding the cooling mechanism of regenerative refrigerators before acoustic heat engines. From a conventional point of view, a regenerator was thought of as a counter flow heat exchanger. From the thermoacoustic point of view, however, the regenerator plays a more active role for making low temperatures by promoting the axial heat flow due to the oscillating gas. Such a difference would

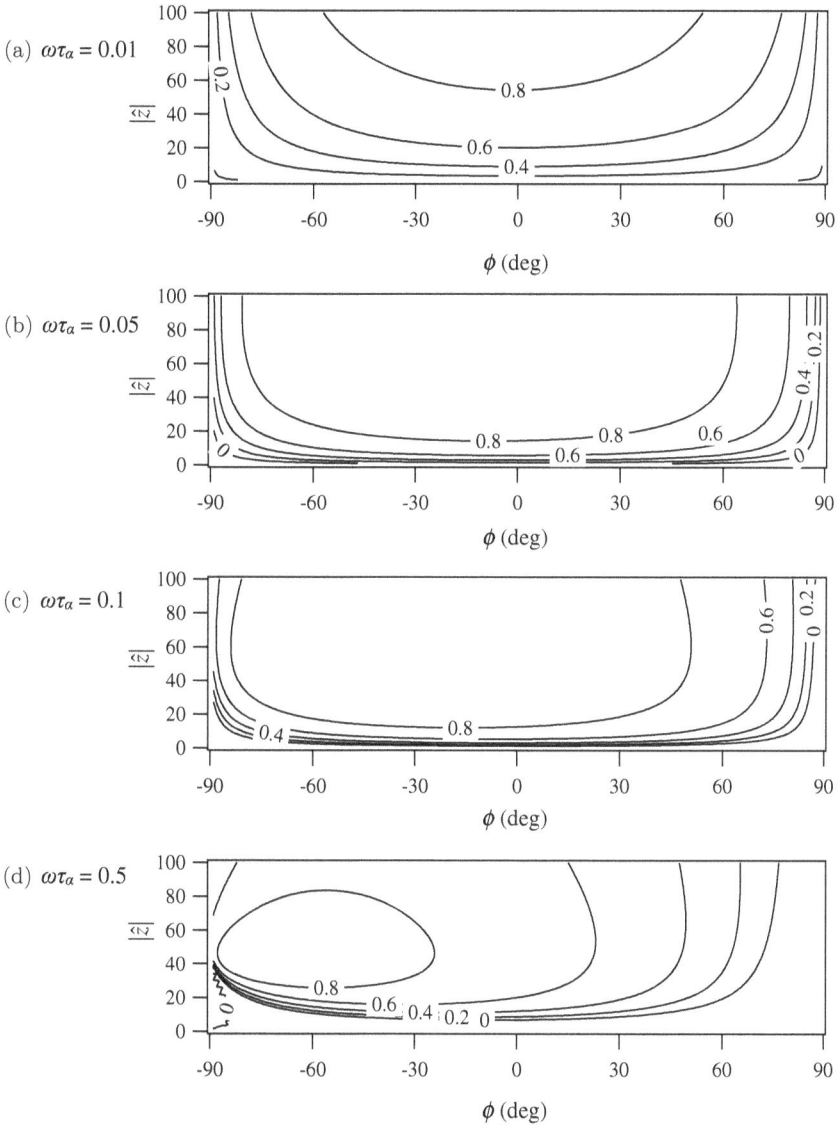

Figure 8.4: Contour plot of ϵ'_C, when $\gamma = 5/3$, $\sigma = 0.66$, and $d\log T_m/dX = -20$; $\hat{z}$ is the dimensionless specific acoustic impedance, whereas ϕ denotes the phase of u_{1r} relative to p_1.

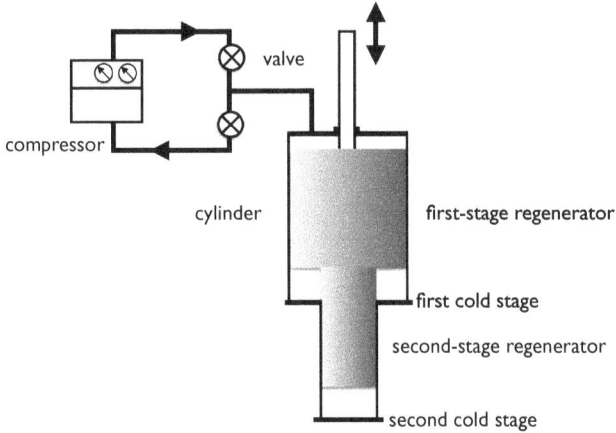

Figure 8.5: Schematic illustration of a GM refrigerator. The working gas is forced to oscillate through regenerators by a reciprocal motion of displacers and also subjected to pressure oscillations by periodic operation of valves connected to a compressor.

not have been easily acceptable to engineers of cryogenics, but the success of pulse tube refrigerators evidenced the applicability of the thermoacoustic theory. Here we discuss the cooling performance of regenerative refrigerators from a thermoacoustic point of view.

A two-stage GM (Gifford-McMahon) refrigerator is schematically shown in Fig. 8.5. The two stage configuration involves two pairs of cylinder and displacer. The second stage reaches about 10 K, while the first stage is also lowered to about 40 K. The regenerator is built in the displacer through which the working gas flows between the expansion space at the bottom and the compressor. The regenerator materials are different in the first-stage regenerator and the second, reflecting the temperature level of respective regenerators: the first-stage regenerator includes a stack of metal screens and the second-stage regenerator contains spherical lead particles. The working gas is usually pressurized helium gas.

Pressure oscillations of the working gas are induced by periodic operation of high-pressure and low-pressure valves connected to a compressor. Also, reciprocal motion of regenerators forces gas to go through the regenerator, which is synchronous to the valve operation.

For brevity, we express the gas displacement $\langle \xi' \rangle_r = |\xi_{1r}| \cos \omega t$ at the cold end of the second-stage regenerator by that of the regenerator, and denote pressure oscillations by $p' = |p_1| \cos(\omega t + \theta)$. In the following, the heat flux density at the cold end of the second stage regenerator is discussed by taking the phase lead θ of pressure relative to displacement, $|p_1|$, $|\xi_{1r}|$ as the control parameters.

When no heat load is supplied to the second cold stage, we can assume that

$$Q_{prog} + Q_{stand} + Q_D = 0 \tag{8.24}$$

by neglecting the contribution from Q_κ. The sum of Q_{prog} and Q_{stand} is given as

$$Q_{prog} + Q_{stand} = -\frac{\omega}{2} \beta T_m |p_1||\xi_{1r}|(\text{Re}[g] \sin \theta + \text{Im}[g] \cos \theta), \tag{8.25}$$

which can be rewritten as

$$Q_{prog} + Q_{stand} = -\frac{\omega}{2} \beta T_m |p_1||\xi_{1r}||g| \sin(\theta + \Theta) \tag{8.26}$$

with

$$\Theta = \arg g. \tag{8.27}$$

Here we consider the critical temperature gradient $(dT_m/dx)_{critical}$ with no heat load condition. Since Q_D is expressed as

$$Q_D = \frac{\omega}{2} \rho_m C_p \left(\frac{dT_m}{dx}\right)_{critical} \text{Im}[g_D]\text{Re}\left[\frac{1}{1 - \chi_\nu}\right]|\xi_{1r}|^2, \tag{8.28}$$

we have

$$\left(\frac{dT_m}{dx}\right)_{critical} = -\frac{\beta T_m}{\rho_m C_p} E \frac{|p_1|}{|\xi_{1r}|} \sin(\theta + \Theta), \tag{8.29}$$

where E means

$$E = \frac{-|g|}{\text{Im}[g_D]\text{Re}\left[\dfrac{1}{1 - \chi_\nu}\right]}. \tag{8.30}$$

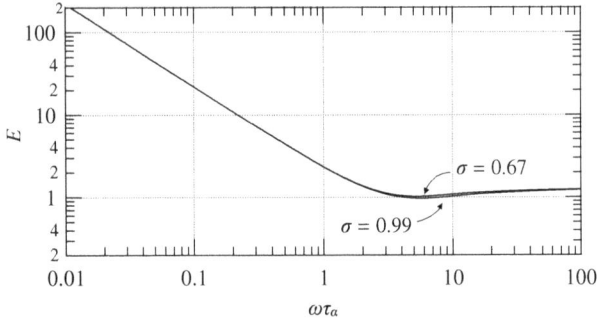

Figure 8.6: $\omega\tau_\alpha$ dependence of E.

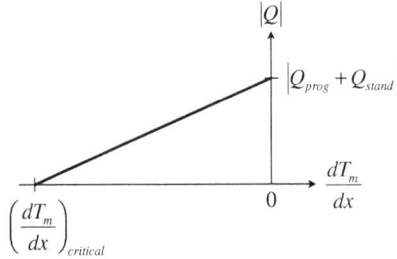

Figure 8.7: Relation between cooling power and temperature gradient.

Figure 8.6 shows $\omega\tau_\alpha$ dependence of E. With decreasing $\omega\tau_\alpha$ below 3, E continues to increase. Therefore, it indicates that we should employ the regenerator with $\omega\tau_\alpha$ as small as possible to increase $(dT_m/dx)_{critical}$ and hence lower the reaching temperature of the refrigerator.

The heat flux density Q can be expressed by using $(dT_m/dx)_{critical}$ as

$$Q = \frac{\omega}{2}\rho_m C_p \mathrm{Im}[g_D]\mathrm{Re}\left[\frac{1}{1-\chi_\nu}\right]|\xi_{1r}|^2\left\{\frac{dT_m}{dx} - \left(\frac{dT_m}{dx}\right)_{critical}\right\}.$$

$$(8.31)$$

Therefore, the cooling power should increase proportional to the temperature gradient dT_m/dx, as shown in Fig. 8.7.

8.3.1 Relation between cooling power and acoustic field (phase lead θ)

The cooling power of a GM refrigerator in Fig. 8.5 is discussed here as a function of a phase lead θ of pressure relative to displacement oscillations at the second cold stage. While Q_D and Q_κ are independent of θ, a sum $Q_{prog} + Q_{stand}$ depends on θ through the relation

$$Q_{prog} + Q_{stand} = -\frac{\omega}{2}\beta T_m |p_1||\xi_{1r}||g|\sin(\theta + \Theta).$$

Thus, the heat flow at the cold stage should be maximized when

$$\theta_{opt} = \frac{\pi}{2} - \Theta \tag{8.32}$$

is achieved.

As shown in Fig. 8.8, $|g|$ and $\Theta = \arg g$ are uniquely determined by $\omega\tau_\alpha$. When $\omega\tau_\alpha \ll 1$, the cooling power should be maximized with $\theta = \pi/2$ since $\Theta = 0$. On the other hand, when $\omega\tau_\alpha \sim \pi$,

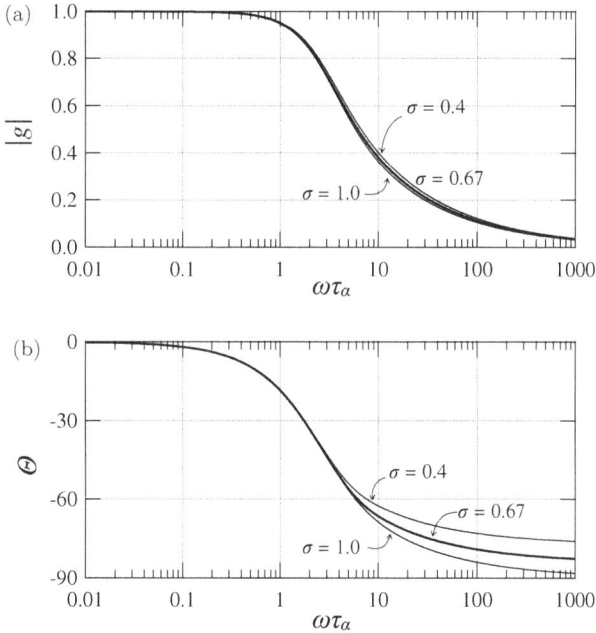

Figure 8.8: Relations between (a) $|g|$ and $\omega\tau_\alpha$ and (b) $\Theta = \arg g$ and $\omega\tau_\alpha$.

it is maximized with $\theta = 3\pi/4$ since $\Theta \approx 45°$; the deviation from the traveling wave phasing ($\theta = \pi/2$) comes from the contribution from Q_{stand} due to irreversible heat exchange processes. It should be noted that $|g|$ is a decreasing function of $\omega\tau_\alpha$. Therefore, in order to increase the heat flow, one should employ the regenerator with $\omega\tau_\alpha$ as low as possible and tune $\theta = \pi/2$.

Figure 8.9 presents the measured cooling power of a GM refrigerator when θ was varied [85]. The first cold stage temperature was kept at 85 K during the measurement of the cooling power at the second stage. As can be seen from Fig. 8.9, the cooling power shows a maximum with $\theta \sim 100°$ when the cooling temperature was as high as 27.5 K. As the cooling temperature was lowered below 12 K, θ goes above 140°. It is known that the thermal diffusivity α of helium

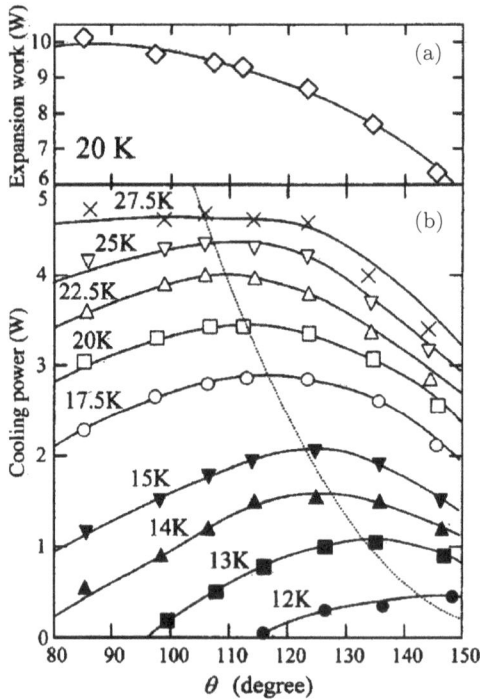

Figure 8.9: Phase lead θ dependence of expansion work and cooling power of a GM refrigerator [85].

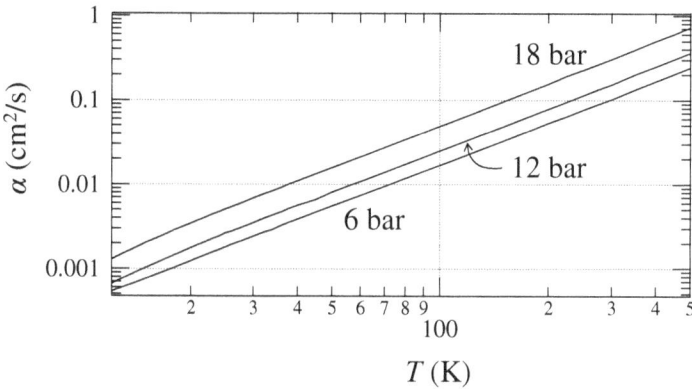

Figure 8.10: Temperature dependence of thermal diffusivity α of helium [86].

decreases drastically with decreasing temperature (see Fig. 8.10). Indeed, α at $10\,\mathrm{K}$ is smaller by an order of magnitude than that at $30\,\mathrm{K}$. Therefore, the thermal relaxation time $\tau_\alpha = r_0^2/(2\alpha)$ becomes greater as the temperature is lowered, which means that the cooling power is enhanced by making θ larger than the traveling wave phasing through the contribution of Q_{stand}.

8.3.2 Relation between cooling power and acoustic field (frequency and amplitude)

The cooling power can be considered as functions of displacement amplitude $|\xi_{1r}|$ and pressure amplitude $|p_1|$. The total heat flux density $Q = Q_{prog} + Q_{stand} + Q_D$ is expressed as

$$Q = -\frac{\omega}{2}\beta T_m |p_1||\xi_{1r}||g|\sin(\theta + \Theta)$$

$$+ \frac{\omega}{2}\rho_m C_p \frac{dT_m}{dx}\mathrm{Im}[g_D]\mathrm{Re}\left[\frac{1}{1 - \chi_\nu}\right]|\xi_{1r}|^2. \qquad (8.33)$$

Therefore, Q increases linearly with increasing $|p_1|$, but Q has a quadratic dependence on $|\xi_{1r}|$. As schematically shown in Fig. 8.11, Q should possess a minimum, namely a maximum cooling power.

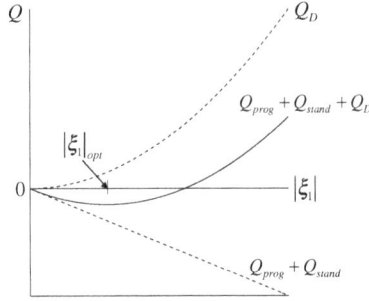

Figure 8.11: Displacement amplitude dependence of heat flux density components.

The optimum $|\xi_{1r}|$ is given by completing the square of Q as

$$|\xi_{1r}|_{opt} = \frac{1}{2} \frac{\beta T_m |g| \sin(\theta + \Theta)}{\rho_m C_p \frac{dT_m}{dx} \text{Im}[g_D] \text{Re}\left[\frac{1}{1 - \chi_\nu}\right]} |p_1|, \qquad (8.34)$$

and the minimum is $(Q_{prog} + Q_{stand})/2$. It is also seen that the cooling power is no longer expected with $|\xi_{1r}| > 2|\xi_{1r}|_{opt}$. This result may be a useful guide when one tries to determine the displacer stroke of a GM refrigerator.

When the cooling power is seen as a function of angular frequency ω, it also shows a maximum, which can be derived as follows. For brevity, let's assume that $\omega \tau_\alpha$ is sufficiently small in the regenerator and inviscid gas. Then set the phase lead θ to the optimum one. Therefore, $\theta_{opt} = \pi/2$ with $\omega \tau_\alpha \ll 1$. The heat flux density can be simplified as

$$Q \approx Q_{prog} + Q_D = -\frac{\omega}{2} \beta T_m \chi'_\alpha |p_1||\xi_1| + \frac{\omega}{2} \chi''_\alpha \rho_m C_p \frac{dT_m}{dx} |\xi_1|^2. \qquad (8.35)$$

When $\omega \tau_\alpha \ll 1$, χ_α is approximated by

$$\chi_\alpha \approx 1 - \frac{(\omega \tau_\alpha)^2}{12} - i \frac{\omega \tau_\alpha}{4}, \qquad (8.36)$$

and hence, we have $\chi'_\alpha \approx 1$ and $\chi''_\alpha \approx -\omega\tau_\alpha/4$. Substituting these relations into Q gives an explicit functional form as a quadratic function of ω. From completing the square, we see that $Q_{prog} + Q_D$ is maximized when

$$\omega_{opt} = -\frac{2}{\tau_\alpha} \frac{\beta T_m}{\rho_m C_p (dT_m/dx)} \left| \frac{p_1}{\xi_1} \right|. \qquad (8.37)$$

As we have seen, lowering the temperature makes the thermal relaxation time τ_α greater and hence ω_{opt} is lowered. In reality, the operation frequency is often tuned to as low as $1\,\mathrm{Hz}$ to $3\,\mathrm{Hz}$ in GM refrigerators.

8.4 Acoustic Field in Pulse Tube Refrigerator

Pulse tube refrigerators use passive flow components like a pulse tube that is an empty tube, buffer tank, and an orifice to control the acoustic field. Simplicity is a remarkable advantage of the pulse tube refrigerators over the other regenerative refrigerators that use mechanically driven displacers and pistons. Here we investigate how the acoustic field is controlled in the pulse tube refrigerator on the basis of the electroacoustic analogy.

8.4.1 *Orifice pulse tube refrigerator*

An orifice pulse tube refrigerator consists of a pulse tube, an orifice, and a tank in addition to a regenerator, heat exchanger, and a driver like a compressor with valve or a reciprocal piston. Figure 8.12(a) schematically shows the basic structure. The working gas occupies the internal volume of the regenerator, pulse tube, and tank. These components are relatively long compared to the gas displacement amplitude, but are much shorter than the wavelength of sound waves. To see the gas pressure and velocity when the driver is in operation, we use the electroacoustic analogy, where pressure p_1 and volume velocity U_1 correspond to voltage and current, respectively.

As we have seen in Section 3.7, the tube segment can be seen as a capacitor or an inductor, depending on the magnitude of

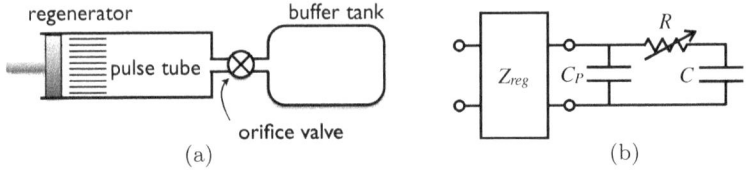

Figure 8.12: (a) Orifice pulse tube refrigerator and (b) its equivalent electric circuit, where C and C_P denote the capacitance of the buffer tank and pulse tube, R means the needle valve; Z_{reg} represents the equivalent impedance of the regenerator.

specific acoustic impedance $|z|$. When $|z|$ is much greater than the characteristic impedance $\rho_m c_S$ of the working gas, the tube segment is seen as the capacitor with capacitance

$$C = Al K_S, \tag{8.38}$$

where A and l are the cross-sectional area and length of the tube segment, and K_S is the adiabatic compressibility. When $|z|$ is much smaller than $\rho_m c_S$, the tube segment is seen as the inductor with inductance

$$L = \frac{\rho_m l}{A}. \tag{8.39}$$

The orifice is assumed by a resistor. In case of a needle valve, it is considered a variable resistor that changes its resistance with opening of the valve. In this way, we can draw an equivalent circuit of the orifice pulse tube refrigerator as shown in Fig. 8.12(b).

Since the heat flow at the cold end of the regenerator reflects the cooling power, we consider the acoustic impedance given by the ratio of complex pressure amplitude p_1 and complex volume velocity amplitude $U_1 = A u_{1r}$ at the cold end

$$Z = \frac{p_1}{U_1} = \frac{1}{A} \left| \frac{p_1}{u_{1r}} \right| e^{-i\phi}. \tag{8.40}$$

It should be noted that ϕ is the phase lead of u_{1r} relative to p_1, and hence the phase angle of Z is $-\phi$. For the equivalent circuit in

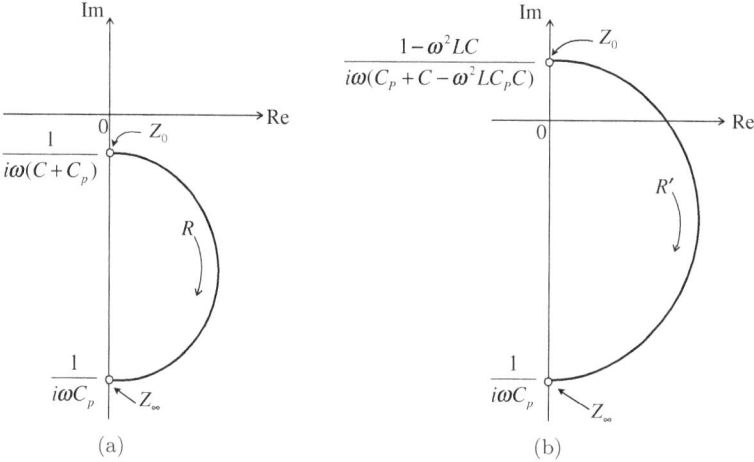

Figure 8.13: Acoustic impedance of (a) orifice pulse tube refrigerator and (b) intertance pulse tube refrigerator. When R and R' are changed from zero to infinity, Z goes from Z_0 to Z_∞.

Fig. 8.12(b), Z is given by

$$\frac{1}{Z} = i\omega C_{\mathrm{P}} + \frac{i\omega C}{1 + i\omega RC} \qquad (8.41)$$

where C and C_{P} respectively denote the capacitance of the buffer tank and pulse tube, and R means the resistance of the needle valve.

The impedance can be adjusted by changing R of the variable resistance. Figure 8.13 shows Z when R is changed from zero to infinity while tracing a semicircle in a complex plane. When R is infinitely large, namely when the valve is closed, Z goes to $Z_\infty = 1/(i\omega C_{\mathrm{P}})$. When R decreases with increasing valve opening, Z approaches $Z_0 = 1/[i\omega(C + C_{\mathrm{P}})]$. The heat flow should change drastically reflecting the valve opening dependence of Z.

The heat flux component $Q_{prog} + Q_{stand}$ is given by

$$Q_{prog} + Q_{stand} = -\frac{1}{2}\beta T_m |p_1||u_{1r}||g|\cos(\phi - \Theta). \qquad (8.42)$$

When the valve is closed, Z_∞ indicates that $\phi = \pi/2$. If $\omega \tau_\alpha$ of the regenerator satisfies $\omega \tau_\alpha \ll 1$, we see $\Theta \approx 0$. Therefore,

$Q_{prog} + Q_{stand}$ results in $Q_{prog} + Q_{stand} = 0$, which explains why the basic pulse tube refrigerator is inferior to the orifice pulse tube refrigerator. When opening the valve, ϕ takes a minimum at a certain valve opening, and then returns to $\pi/2$. Therefore, the orifice pulse tube refrigerator should maximize the cooling performance when the minimum ϕ is achieved. The minimum valve opening comes close to a traveling wave phasing but never becomes negative, if the buffer tank volume is increased. However, with the infinitely large buffer tank, the impedance $|Z|$ becomes infinitely small with $\phi = 0$. Then the heat loss would increase. Therefore, there should be an optimum size of the buffer tank, as well as optimum valve opening in the orifice valve.

8.4.2 *Inertance pulse tube refrigerator*

As indicated from $Q_{prog} + Q_{stand}$, the cooling power should become greatest with $\phi = \Theta$, where $-\pi/2 < \Theta < 0$. Therefore, the cooling power should be improved if one can make ϕ negative. Use of the inertance tube offers one solution to realize negative ϕ.

Figure 8.14(a) presents an inertance tube pulse tube refrigerator, which has an inertance tube between a needle valve and a buffer tank. The acoustic impedance is given from the equivalent circuit where the inertance tube is replaced with a series of coil with inductance L and a resistance R_ν by

$$\frac{1}{Z} = i\omega C_{\mathrm{P}} + \frac{i\omega C}{1 - \omega^2 LC + i\omega R'C} \qquad (8.43)$$

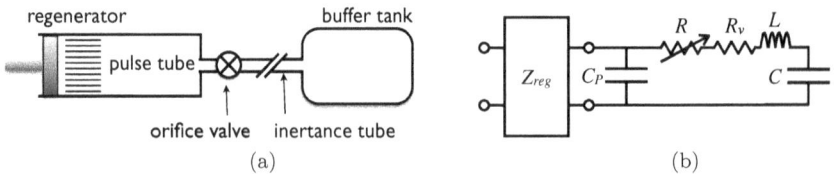

Figure 8.14: Schematic illustration of (a) inertance tube pulse tube refrigerator and (b) the equivalent circuit.

with $R' = R + R_\nu$. When R' is infinitely large, Z goes to $Z_\infty = 1/(i\omega C_P)$; when $R' = 0$, Z reaches

$$Z_0 = \frac{1 - \omega^2 LC}{i\omega(C_P + C - \omega^2 LC_P C)}. \tag{8.44}$$

Therefore, if the imaginary part of Z_0 is positive, ϕ should become negative above a certain value of R' as in Fig. 8.13(b). The condition for $\mathrm{Im}\, Z > 0$ is given by

$$\frac{1}{\omega C} < \omega L < \frac{1}{\omega C_P} + \frac{1}{\omega C}, \tag{8.45}$$

which means that $\phi < 0$ is not achieved if L of the inertance tube is too large.

The pulse tube refrigerator performance was greatly improved by the introduction of the orifice and the inertance tube. Inoue, who was an industry engineer responsible for the research and development of pulse tube refrigerators, said "It became a completely different refrigerator" when he opened the needle valve in his first experiment with the orifice pulse tube refrigerator. A drastic decrease of the cooling temperature by a small change of the valve opening must have been a big surprise.

8.4.3 *Experimental verification of passive acoustic field controllers*

Acoustic field control by the orifice and inertance tube has been demonstrated experimentally [87]. Figure 8.15 shows a prototype of double-inlet pulse tube refirigerator. The regenerator is made of stainless steel mesh screens with mesh number 250 and wire diameter 0.04 mm, stacked in a thin-walled cylinder 50 mm in length and 14 mm in diameter. On top of the regenerator, the cold heat exchanger and the pulse tube made of glass were placed. The buffer tank was connected to the end of the pulse tube via the needle valve. Another needle valve, called a bypass valve, was on the small pipe connecting the top of the pulse tube and the inlet of the regenerator. When both the needle valve and the bypass valve are closed, the refrigerator operates as a basic pulse tube refrigerator. When the

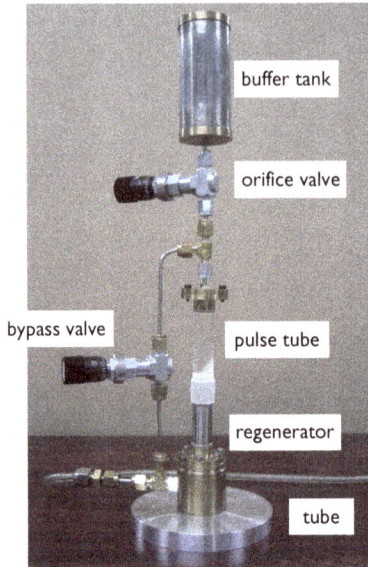

Figure 8.15: Experimental prototype of a double-inlet pulse tube refrigerator.

needle valve on the top of the pulse was opened, it works as a orifice pulse tube refrigerator; when the bypass valve is also open, it becomes a double-inlet pulse tube refrigerator. When only the bypass valve is open, it is the inertance tube pulse tube refrigerator. Reference [87] reports the acoustic impedance determined by simultaneous measurements of pressure and velocity, which gave experimental support to the analysis based on the equivalent circuit.

8.5 Dream Pipe

The heat flux density component Q_D is essential to the operation of dream pipes that promote heat transfer from hot to cold through oscillating motion of fluids. Let us briefly recall the basic physics of Q_D. Consider a fluid parcel oscillating along a flow channel that has an axial temperature gradient. The fluid temperature goes up and down along with the displacement oscillation due to change of the local wall temperature. When the temperature gradient and the displacement amplitude are denoted by dT_m/dx and $|\xi|$, the maximum temperature difference from the local wall would be roughly expressed as $(dT_m/dx)|\xi|$. The heat dq exchanged between the fluid and the wall is then given by $\rho_m C_p(dT_m/dx)|\xi|$, where ρ_m and C_p are the temporal mean density and isobaric specific heat of the fluid. Since the fluid absorbs dq at its hot end and releases it at its cold end, the heat transfer by the fluid parcel is $dq \times |\xi|$ per acoustic period. Therefore, Q_D is approximately given by $\omega \rho_m C_p(dT_m/dx)|\xi|^2$. In contrast to Q_{prog} and Q_{stand}, the thermal expansion coefficient β is not involved in Q_D. Instead, Q_D is proportional to the heat capacity $\rho_m C_p$ per volume. Thus, Q_D is more enhanced in fluids than in gases.

A more general treatment in Chapter 6 gives Q_D as

$$Q_D = \frac{\omega}{2} \rho_m C_p \frac{dT_m}{dx} \text{Im}[g_D] \text{Re}\left[\frac{1}{1-\chi_\nu}\right] |\xi_{1r}|^2. \qquad (8.46)$$

Figure 8.16 illustrates $\text{Im}[g_D]$ as a function of $\omega\tau_\alpha$, where we see that $\text{Im}[g_D]$ shows minimum near $\omega\tau_\alpha \approx 2.8$, indicating the importance of irreversible heat exchange processes. It should be also noted that $\text{Im}[g_D]$ is always negative, and hence Q_D has an opposite sign to dT_m/dx. Therefore, Q_D always flows from hot to cold, in the same way as the conduction heat.

Assuming water as the working fluid, discuss here Q_D quantitatively. We introduce the effective thermal conductivity by

$$Q_D = -\kappa_D \frac{dT_m}{dx} \qquad (8.47)$$

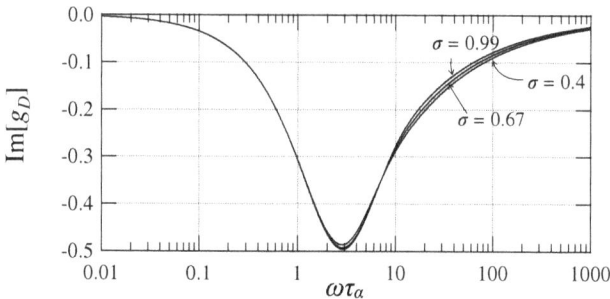

Figure 8.16: Relation between $\text{Im}[g_D]$ and $\omega\tau_\alpha$.

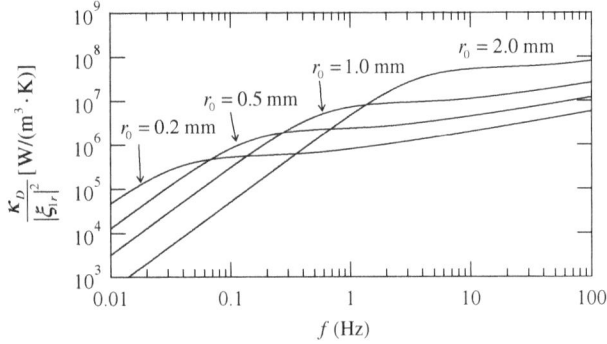

Figure 8.17: Relation between $\kappa_D/|\xi_{1r}|^2$ and f; κ_D is the effective thermal conductivity, $|\xi_{1r}|$ is the amplitude of cross-sectional mean displacement oscillation, f is the frequency. Water of 300 K at ambient pressure is considered as a working fluid. r_0 in figure denotes the pipe radius assumed.

with

$$\kappa_D = -\frac{\omega}{2}\rho_m C_p \text{Im}[g_D]\text{Re}\left[\frac{1}{1-\chi_\nu}\right]|\xi_{1r}|^2. \qquad (8.48)$$

Figure 8.17 shows the ratio $\kappa_D/|\xi_{1r}|^2$

$$\frac{\kappa_D}{|\xi_{1r}|^2} = -\frac{\omega}{2}\rho_m C_p \text{Im}[g_D]\text{Re}\left[\frac{1}{1-\chi_\nu}\right] \qquad (8.49)$$

as a function of oscillation frequency f for water at 300 K and ambient pressure [Prandtl number $\sigma = 5.8$, $\rho_m = 997\,\text{kg/m}^3$, $C_p = 4.18\,\text{kJ/(kg·K)}$], when tube radius r_0 is varied from 0.2 mm to 2.0 mm. Broad humps in curves of different r_0 values reflect the minimum of $\text{Im}[g_D]$ in Fig. 8.17. When $r_0 = 0.5$ mm, $\kappa_D/|\xi_{1r}|^2$ reaches $7.3 \times 10^6\,\text{W/(m}^3\text{·K)}$ with $f = 1.0$ Hz. Therefore, the effective thermal conductivity κ_D can exceed that of Cu [$\kappa \approx 400\,\text{W/(m·K)}$] when the displacement amplitude is 7.4 mm. Furthermore, κ_D goes to $5300\,\text{W/(m·K)}$ with $r_0 = 0.2$ mm, $|\xi_{1r}| = 10$ mm, and $f = 10$ Hz. This thermal conductivity is 13 times larger than that of Cu, and is 900 times larger than that of still water [$\kappa \approx 0.61\,\text{W/(m·K)}$]. Since the thermal conductivity is widely controllable through oscillation frequency and amplitude, the dream pipe would be useful as a heat switch as well as a heat conduction device.

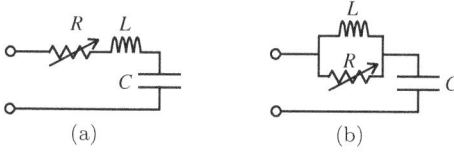

Figure 8.18: Equivalent electrical circuits of the considered pulse tube refrigerators; R: (variable) flow resistance of the orifice valve, L: inductance of the inertance tube, and C: capacitance of the reservoir.

8.6 Problems

1 Consider pulse tube refrigerators whose pulse tube ends are terminated by an orifice valve, inertance tube, and reservoir. When the equivalent electrical circuits are illustrated as shown in Figs. 8.18(a) and (b), answer the questions.

(1) Derive the impedance Z of the terminator.
(2) Show the trace of Z on the complex plane when R is changed.

Answers

1 (1) The impedances Z_a and Z_b of the circuits (a) and (b) are respectively given by

$$Z_a = R + i\omega L + \frac{1}{i\omega C}$$

and

$$Z_b = \frac{i\omega L}{1 + i\omega L/R} + \frac{1}{i\omega C}.$$

(2) See Figs. 8.19(a) and 8.19(b).
Hint: the trace of $Z = i\omega L/(1 + i\omega L/R)$ can be considered as follows. By introducing a variable $x = \omega L/R$, $z = Z/(\omega L)$ is transformed as

$$z = \frac{i}{1 + ix} = \frac{x + i}{1 + x^2}.$$

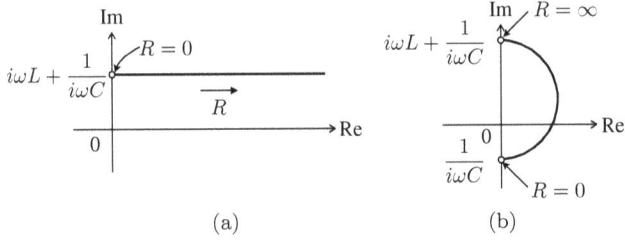

Figure 8.19: Trace of the impedance Z: (a) the real part of Z increases with increasing R, with keeping the imaginary part at $i\omega L + 1/(i\omega C)$, (b) Z draws a semicircle with diameter ωL. Note the sign of $i\omega L + 1/(i\omega C)$ is positive when $\omega > \omega_0$ and negative when $\omega < \omega_0$, where $\omega_0 = \sqrt{1/(LC)}$.

For $z = z' + iz''$, we have

$$z' = \frac{x}{1+x^2}, \quad z'' = \frac{1}{1+x^2}.$$

By eliminating x from the coupled equations above, we obtain

$$z'^2 + \left(z'' - \frac{1}{2}\right)^2 = \frac{1}{4}.$$

This equation represents the circle centered at $(0, 1/2)$, with diameter of 1.

Chapter 9

Future Prospects

Research and development of thermoacoustic devices are in progress over the world, on the basis of thermoacoustic theory and relevant experimental verifications. Recent studies, however, have faced problems that require further understanding of various aspects of oscillating flow dynamics. In this chapter, some of them are introduced and future prospects are given.

9.1 Towards Practical Applications of Thermoacoustic Devices

For the development of practical thermoacoustic engines, it is necessary to establish a design method. Numerical simulation of thermal and flow behaviors should be one of the necessary tools. Here we introduce the principle of the calculation method that fully makes use of thermoacoustic theory, and also the recent research activities on heat exchangers.

9.1.1 *Calculation method based on thermoacoustic theory*

Equation of motion for viscous fluid and continuity equation expressed by using the solution of a general equation of heat transfer for thermoviscous fluid were introduced in Chapters 3 and 5, respectively. When the volumetric velocity $U_1 = Au_{1r}$ is used in place of the cross-sectional mean velocity u_{1r} (A is the cross-sectional area of the flow channel), those equations are given by

$$\frac{dp_1}{dx} = -\frac{i\omega\rho_m}{A(1 - \chi_\nu)}U_1 \tag{9.1}$$

$$\frac{dU_1}{dx} = -i\omega AK_E p_1 + \beta_E \frac{dT_m}{dx}U_1 \tag{9.2}$$

where K_E and β_E are the effective compressibility and effective thermal expansion coefficient

$$K_E = K_S + (K_T - K_S)\chi_\alpha \tag{9.3}$$

$$\beta_E = \beta\langle b\rangle_r \quad \text{with} \quad \langle b\rangle_r = \frac{\chi_\alpha - \chi_\nu}{(1 - \chi_\nu)(1 - \sigma)}. \tag{9.4}$$

For an ideal gas, K_E and β_E are given by

$$K_E = \frac{1}{\gamma p_m}\{\, 1 + (\gamma - 1)\chi_\alpha\} \tag{9.5}$$

$$\beta_E = \frac{1}{T_m}\frac{\chi_\alpha - \chi_\nu}{(1 - \chi_\nu)(1 - \sigma)}. \tag{9.6}$$

Therefore, Eqs. (9.1) and (9.2) are written in a matrix form as

$$\frac{d}{dx}\begin{pmatrix} p_1(x) \\ U_1(x) \end{pmatrix} = \begin{pmatrix} 0 & m_{12} \\ m_{21} & m_{22} \end{pmatrix}\begin{pmatrix} p_1(x) \\ U_1(x) \end{pmatrix}, \tag{9.7}$$

where

$$m_{12} = -\frac{i\omega\rho_m}{A(1 - \chi_\nu)} \tag{9.8}$$

$$m_{21} = -\frac{i\omega A\{\, 1 + (\gamma - 1)\chi_\alpha\}}{\gamma p_m} \tag{9.9}$$

$$m_{22} = \frac{\chi_\alpha - \chi_\nu}{(1 - \chi_\nu)(1 - \sigma)}\frac{1}{T_m}\frac{dT_m}{dx}. \tag{9.10}$$

When a temperature gradient exists along x, the matrix components m_{12}, m_{12}, and m_{22} depend on x through temperature dependence of thermal properties of the fluid. Therefore, one needs to numerically

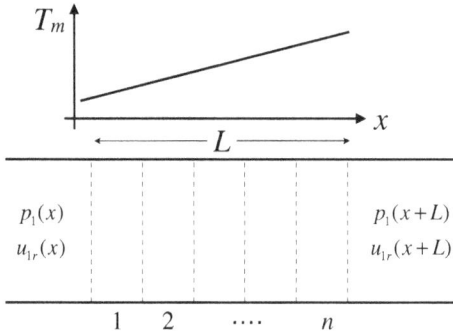

Figure 9.1: Axial temperature distribution along a flow channel of length L. The flow channel is divided into n-short segments.

integrate Eq. (9.7). However, if we consider a sufficiently short flow segment, temperature dependent quantities are approximated by those at the mean temperature, and also the temperature gradient is seen as a constant. More specifically, we divide a flow channel of length L into n segments, as shown in Fig. 9.1, and solve Eq. (9.7) for each segment with length l analytically following a prescription of ordinary differential equations. The solution for each segment is given by using a transfer matrix

$$M_j = e^{m_{22}l/2} \begin{pmatrix} -\dfrac{m_{22}}{D} \sinh\left(\frac{Dl}{2}\right) + \cosh\left(\frac{Dl}{2}\right) & -\dfrac{2m_{12}}{D} \sinh\left(\frac{Dl}{2}\right) \\ -\dfrac{2m_{21}}{D} \sinh\left(\frac{Dl}{2}\right) & \dfrac{m_{22}}{D} \sinh\left(\frac{Dl}{2}\right) + \cosh\left(\frac{Dl}{2}\right) \end{pmatrix}$$

$$\tag{9.11}$$

$$D = \sqrt{m_{22}^2 + 4m_{12}m_{21}} \tag{9.12}$$

and the pressure $p_1(x + l)$ and volume velocity $U_1(x + l)$ at position $x + l$ are linked with $p_1(x)$ and $U_1(x)$ at position x by

$$\begin{pmatrix} p_1(x + l) \\ U_1(x + l) \end{pmatrix} = M_j \begin{pmatrix} p_1(x) \\ U_1(x) \end{pmatrix}. \tag{9.13}$$

By successively multiplying the transfer matrices $M_1, M_2, \ldots, M_n$, of n segments we have the following relation between $(p_1(x+L), U_1(x+L))$ and $(p_1(x), U_1(x))$

$$\begin{pmatrix} p_1(x+L) \\ U_1(x+L) \end{pmatrix} = M \begin{pmatrix} p_1(x) \\ U_1(x) \end{pmatrix} \tag{9.14}$$

$$M = M_1 M_2 \cdots M_n. \tag{9.15}$$

Thermoacoustic devices consist of various flow channels like regenerator, heat exchanger, and resonance tube. Each flow channel is represented by the transfer matrix whose matrix components reflect the channel geometry and channel wall temperature. For example, consider a thermoacoustic system in Fig. 9.2, which consists of a tube, cold heat exchanger, regenerator, thermal buffer tube, hot heat exchanger, thermal buffer tube, and tube. The transfer matrix M_{TB} for the thermal buffer tube is represented by M given above. The transfer matrix M_{reg} for the regenerator is also expressed by M if it is approximated by a bundle of tubes with radius r_0. In that case, the cross-sectional area A should be taken as $A = \epsilon \pi r_0^2$, where ϵ denotes the porosity. The resonance tube (length l_{tub}) possesses uniform temperature. Therefore, $m_{22} = 0$ and the transfer matrices reduces to

$$M_{tub} = \begin{pmatrix} \cos k l_{tub} & -iZ \sin k l_{tub} \\ \dfrac{\sin k l_{tub}}{iZ} & \cos k l_{tub} \end{pmatrix} \tag{9.16}$$

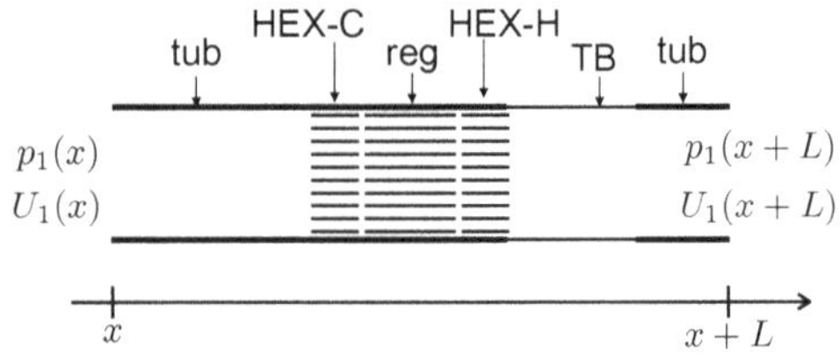

Figure 9.2: Schematic illustration of a thermoacoustic system; tub: resonance tube with uniform temperature, HEX-C: cold heat exchanger, reg: regenerator, HEX-H: hot heat exchanger, TB: thermal buffer tube.

where

$$k = \frac{\omega}{c_S} \sqrt{\frac{1 + (\gamma - 1)\chi_\alpha}{1 - \chi_\nu}} \qquad (9.17)$$

$$Z = \frac{\omega \rho_m}{Ak(1 - \chi_\nu)}. \qquad (9.18)$$

If the heat exchanger is assumed to have uniform temperature, the transfer matrix M_{HEX} is also expressed by the same transfer matrix as the resonance tube, but with elevated or degraded temperature. If the heat exchanger is made of parallel plates with spacing $2r_0$, we should employ

$$\chi_j = \frac{\tanh\left\{(i+1)\dfrac{r_0}{\delta_j}\right\}}{(i+1)\dfrac{r_0}{\delta_j}}, \quad (j = \nu, \alpha). \qquad (9.19)$$

In this way, the total transfer matrix for the system shown in Fig. 9.2 is expressed by

$$M_{total} = M_{tub} M_{TB} M_{HEX,H} M_{reg} M_{HEX,C} M_{tub} \qquad (9.20)$$

and the pressure and volume velocity at ends are related to each other through the relation

$$\begin{pmatrix} p_1(x + L) \\ U_1(x + L) \end{pmatrix} = M_{total} \begin{pmatrix} p_1(x) \\ U_1(x) \end{pmatrix}. \qquad (9.21)$$

By using p_1 and U_1 obtained from Eq. (9.21), one would be able to calculate the axial distribution of p_1 and U_1, from which one can determine the work flow $\tilde{I} = (1/2)\text{Re}[p_1 U_1^\dagger]$, and the complex acoustic impedance p_1/U_1.

For a description of p_1 and U_1 in a given thermoacoustic device, we need to consider the boundary conditions. Some examples are shown in the following.

Resonance tube acoustic engine

Consider a resonance tube acoustic engine with both ends closed as shown in Fig. 9.3(a). We are able to set $U_1 = 0$ at both ends of the resonator. Therefore, the pressure $p_1(0)$ and $p_1(L)$ at both ends satisfy

$$\begin{pmatrix} p_1(L) \\ 0 \end{pmatrix} = M_{total} \begin{pmatrix} p_1(0) \\ 0 \end{pmatrix}.$$ (9.22)

In order to have non-zero $p_1(0)$ and $p_1(L)$, the transfer matrix component $(M_{total})_{2,1}$ is zero

$$(M_{total})_{2,1} = 0.$$ (9.23)

From this equation, one can obtain the frequency and the temperature difference of the resonance tube acoustic engine.

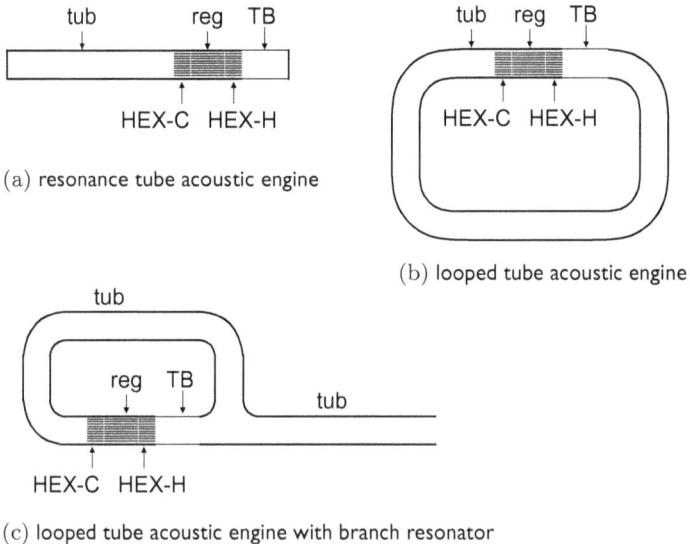

(a) resonance tube acoustic engine

(b) looped tube acoustic engine

(c) looped tube acoustic engine with branch resonator

Figure 9.3: Schematic illustration of acoustic engines: (a) resonance tube acoustic engine, (b) looped tube acoustic engine, (c) looped tube engine with branch resonator.

Looped tube acoustic engine

Consider a looped tube engine shown in Fig. 9.3(b). Because the periodic boundary condition is applied, we have to have a relation

$$\begin{pmatrix} p_1(x_0) \\ U_1(x_0) \end{pmatrix} = M_{total} \begin{pmatrix} p_1(x_0) \\ U_1(x_0) \end{pmatrix}, \qquad (9.24)$$

where x_0 is taken somewhere in the tube region, and x is directed in the clockwise direction. For non-zero $p_1(x_0)$ and $U_1(x_0)$, the determinant of the matrix $M_{total} - E$ must be zero

$$\{(M_{total})_{1,1} - 1\}\{(M_{total})_{2,2} - 1\} - (M_{total})_{1,2}(M_{total})_{2,1} = 0 \qquad (9.25)$$

where E is identity matrix. From Eq. (9.25), one can obtain the frequency and the temperature difference of the looped tube acoustic engine.

Looped tube acoustic engine with branch resonator

Consider a looped tube engine shown in Fig. 9.3(c). If we take x in the anti-clockwise direction and set x_0 at the connecting point of the loop and the resonator, we have

$$\begin{pmatrix} p_1(x_0) \\ U_1(x_0) \end{pmatrix} = M_{total} \begin{pmatrix} p_1(x_0)' \\ U_1(x_L) \end{pmatrix}, \qquad (9.26)$$

where x_L denotes the same position as x_0, and also we have used the relation $p_1(x_0) = p_1(x_L)$. The volume velocity flowing from the loop to the resonator is written as $U_1(x_L) - U_1(x_0)$. If the resonator end $(x = x_R)$ is open and p_1 is zero, we have

$$\begin{pmatrix} 0 \\ U_1(x_R) \end{pmatrix} = M_{tub} \begin{pmatrix} p_1(x_0) \\ U_1(x_L) - U_1(x_0) \end{pmatrix}. \qquad (9.27)$$

By combining Eqs. (9.26) and (9.27), we have

$$\begin{pmatrix} p_1(x_0) \\ U_1(x_0) \end{pmatrix} = M_{all} \begin{pmatrix} p_1(x_0) \\ U_1(x_0) \end{pmatrix}, \qquad (9.28)$$

where

$$M_{all} = M_{total} \begin{pmatrix} 1 & 0 \\ -\dfrac{(M_{tub})_{11}}{(M_{tub})_{12}} & 1 \end{pmatrix}. \qquad (9.29)$$

Therefore, we have

$$\{(M_{all})_{1,1} - 1\}\{(M_{all})_{2,2} - 1\} - (M_{all})_{1,2}(M_{all})_{2,1} = 0 \qquad (9.30)$$

from which we can determine the frequency and the temperature.

Several papers [40, 68, 77, 82, 88, 89] demonstrated that the numerically solved frequency and temperature difference gave a good agreement with the experiments. Therefore, the calculation method based on the thermoacoustic theory should provide a guide to design thermoacoustic devices. For more practical design issues, one would need to know losses due to minor loss and acoustic streaming [90–95], and also how to treat the flow channels in stacked mesh screens [96,97].

9.1.2 *Heat exchangers in oscillatory flow*

Heat exchangers provide a thermal link between the working fluid and the external heat source/sink to promote heat transfer between them. Whereas the net heat transfer between the fluid and the wall is zero over an oscillation period in the regenerator region, the net heat transfer is non-zero in the fluid and heat exchanger wall. Otherwise, no heat input and heat release take place in the thermoacoustic device.

If we consider a cylindrical tube as a flow channel of the heat exchanger, the net heat transfer rate $\bar{q}$ to the fluid from the wall per unit time is given by

$$\bar{q} = \kappa \left\langle \left. \frac{d(T_m + T')}{dr} \right|_{r_0} \right\rangle_t = \kappa \left. \frac{dT_m}{dr} \right|_{r_0} \qquad (9.31)$$

where T_m denotes the temporal mean temperature and r_0 is the tube radius; κ is the thermal conductivity of the fluid. If $dT_m/dr > 0$, the net heat transfer takes place from the wall to the fluid, while if $dT_m/dr < 0$, then the net heat transfer occurs in the opposite direction. When $dT_m/dr = 0$, as assumed in the thermoacoustic theory, no net heat transfer occurs.

Heat exchangers are required to attain/release the larger $\bar{q}$ with the smaller difference ΔT_m of mean temperatures between the gas and the channel walls while maintaining small viscous losses. In other words, the heat transfer coefficient h given by

$$h = \frac{\bar{q}}{\Delta T_m} \tag{9.32}$$

should be as high as possible with as low viscous damping as possible. If ΔT_m becomes large, the deviation of the regenerator end temperatures from the heat source/sink temperatures also becomes large. Therefore, the Carnot efficiency, which is determined by the end temperatures of the regenerator, decreases inevitably, even if the source/sink temperatures are kept constant.

The heat transfer coefficient h has been an interesting subject in the thermoacoustic community [17, 98, 99]. Swift [100] suggested

$$h = \frac{\kappa}{\delta_\alpha}, \tag{9.33}$$

where $\delta_\alpha = \sqrt{2\alpha/\omega}$ is the thermal boundary layer thickness. Subsequently, Mozurkewich [101, 102] presented a similar expression $h \sim 0.61\kappa/\delta_\alpha$ from his theoretical analysis. Kobayashi et al. [103] measured the time-averaged temperature T_m in a cylindrical channel, and observed a parabolic temperature profile in the cross-section of the tube. Then he deduced the heat transfer coefficient

$$h = \frac{2\kappa}{r_0}. \tag{9.34}$$

These studies indicate that h is governed by the thermal conductivity κ of the fluid and the characteristic transverse lengths δ_α and r_0, but not by flow velocity amplitude. If h can be treated as a constant, we can address the problem of how large the heat exchanger should be for a target output power or thermal efficiency in the following way.

Relation between heat exchanger and engine performance

We consider a thermoacoustic engine schematically shown in Fig. 9.4, where two heat exchangers and a regenerator are built in a pipe filled with a working gas [121]. Suppose that the hot heat exchanger is in thermal contact with a heat source of temperature T_H and the cold heat exchanger is in thermal contact with a heat sink of temperature T_C. Heat transfer between the heat exchanger and heat source/sink is assumed to obey Newton's law. Namely, the heat transfer rate per unit time, Q_H, entering the regenerator from the heat source through the hot heat exchanger is written as

$$Q_H = (hA)_H(T_H - T'_H) \tag{9.35}$$

where T'_H is the characteristic gas temperature in the hot heat exchanger and $(hA)_H$ denotes the product of the heat transfer coefficient h and the heat transfer area A of the hot heat exchanger. Also, we have

$$Q_C = (hA)_C(T'_C - T_C) \tag{9.36}$$

where Q_C is the heat transfer rate rejected from the cold side of the regenerator to the heat sink via the cold heat exchanger; T'_C is the characteristic gas temperature in the cold heat exchanger, $(hA)_C$ is the product of h and A of the cold heat exchanger.

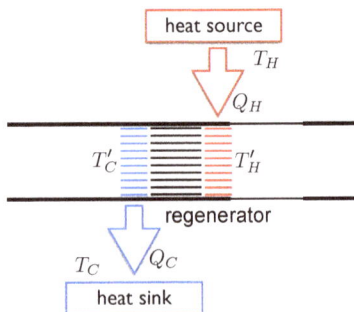

Figure 9.4: Acoustic engine model. Heat Q_H from heat source with temperature T_H goes into the regenerator from the hot end with temperature T'_H, while Q_C leaves the regenerator cold end with T'_C to heat sink with T_C.

The thermoacoustic energy conversion in the regenerator can be discussed in the framework of thermoacoustic theory, as we have done in Chapter 7. The theory formulates thermal efficiency as

$$\eta = \frac{W}{Q_H} = \epsilon \eta_{Carnot} \quad (\epsilon < 1) \tag{9.37}$$

where W, η_{Carnot}, and ϵ are the output power, Carnot efficiency, and a constant given by the acoustic field and regenerator characteristics. The Carnot efficiency is expressed by

$$\eta_{Carnot} = 1 - \frac{T'_C}{T'_H}. \tag{9.38}$$

The first law of thermodynamics states

$$W = Q_H - Q_C. \tag{9.39}$$

Therefore,

$$1 - \frac{Q_C}{Q_H} = \epsilon \left(1 - \frac{T'_C}{T'_H} \right). \tag{9.40}$$

By Eqs. (9.35) and (9.36), Eq. (9.40) can be transformed as

$$1 - \frac{(hA)_C}{(hA)_H} \frac{\frac{T'_C}{T'_H} \frac{T'_H}{T_H} - \frac{T_C}{T_H}}{1 - \frac{T'_H}{T_H}} = \epsilon \left(1 - \frac{T'_C}{T'_H} \right). \tag{9.41}$$

For brevity, we introduce the following dimensionless constants;

$$c = \frac{(hA)_C}{(hA)_H}, \quad \theta = \frac{T_C}{T_H} \tag{9.42}$$

and variables

$$x = \frac{T'_C}{T'_H}, \quad y = \frac{T'_H}{T_H}. \tag{9.43}$$

Then Eq. (9.43) is rewritten as

$$1 - c\frac{xy - \theta}{1 - y} = \epsilon(1 - x). \tag{9.44}$$

If we solve the equation with respect to y, we have

$$y(x) = \frac{\epsilon - c\theta - 1 - \epsilon x}{\epsilon - 1 - (c + \epsilon)x}. \tag{9.45}$$

The output power W is expressed by x and y as

$$W = \epsilon(1 - x)Q_H \tag{9.46}$$

$$= \epsilon(hA)_H T_H(1 - x)(1 - y). \tag{9.47}$$

Finally, we obtain

$$W = \epsilon(hA)_H T_H(1 - x)\frac{c(\theta - x)}{\epsilon - 1 - (c + \epsilon)x}, \tag{9.48}$$

and then, Q_H is given by

$$Q_H = (hA)_H T_H \frac{c(\theta - x)}{\epsilon - 1 - (c + \epsilon)x}. \tag{9.49}$$

Now we have mathematical expression for W, Q_H, T_H', T_C', and Q_C as a function of x, where $T_H' = yT_H$, $T_C' = xyT_H$, $Q_C = Q_H - W$. In other words, these quantities are parametrically linked by $x = T_C'/T_H'$. In the following, we present the relation between $T_{H,C}'$, W, and Q_C as functions of Q_H, after evaluating these quantities as a function of x.

As an example, let us show you the case with $(hA)_H = (hA)_C = 7$ W/K, $\epsilon = 0.9$, $T_H = 150°C$, $T_C = 23°C$. Figure 9.5(a) presents the output power W as a function of Q_H. When $Q_H = 0$, the output power W is zero, and the temperatures of T_H' and T_C' are equal to those of heat source and sink, as shown in Fig. 9.5(b). Owing to the largest temperature difference between T_H' and T_C', the thermal efficiency $\eta = W/Q_H$, as shown in Fig. 9.5(c), is the highest at infinitesimally small Q_H. As Q_H is increased, W also increases but starts to decrease after showing a peak at $Q_H = 245$ W. The reduction is caused by the reduction of Carnot efficiency, since two temperatures of T_H' and T_C' become close to each other as Q_H is increased. Finally, when two temperatures coincide with each other, the output power becomes zero with $Q_H = 445$ W. Although the thermal efficiency η is a decreasing function of Q_H, the output power

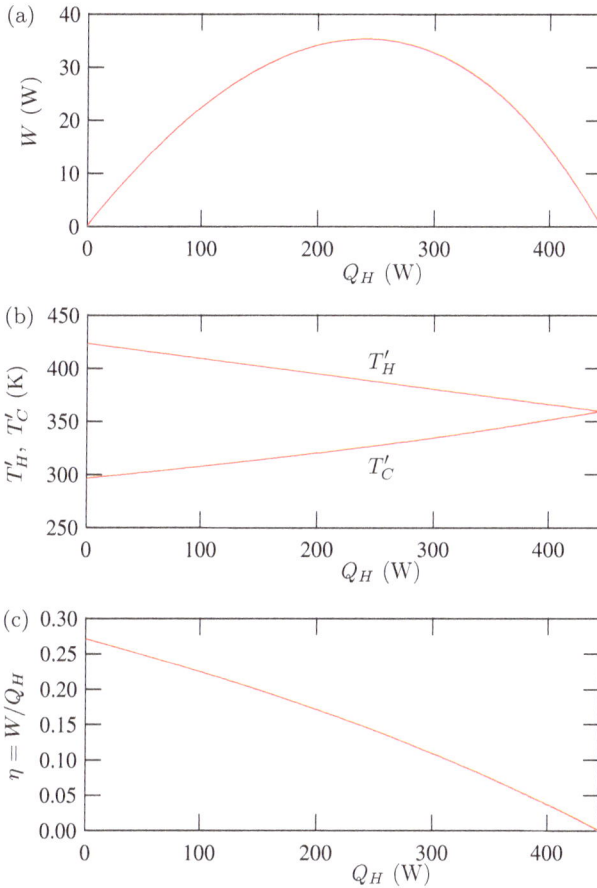

Figure 9.5: (a) Output power W, (b) temperatures $T'_{H,C}$ and (c) efficiency η as a function of Q_H.

becomes the maximum at $Q_H = 245$ W. Therefore, there is no gain both in efficiency and output if Q_H is increased more than that.

Figure 9.6 shows W versus Q_H curves for $hA = 7, 10$, and 13 W/K when $c = 1$, $\epsilon = 0.9$, $T_H = 150°$, and $T_C = 23°$C. The maximum value, W_{max}, of the output power increases with hA and also the associated heat power $Q_{H,max}$ increases: when hA is equal to 7 W/K, W_{max} does not reach 50 W, whereas when $hA = 10$ W/K, W_{max} reaches 50 W at $Q_{H,max} = 344$ W and efficiency is 14%. When hA is

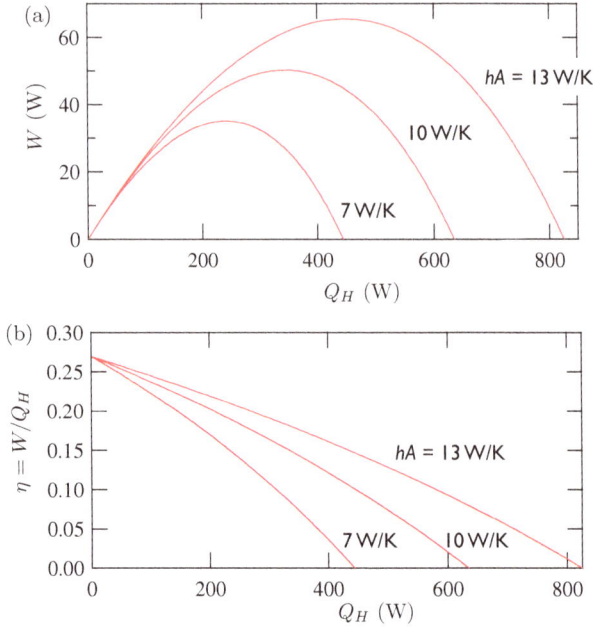

Figure 9.6: Output power W versus Q_H when $hA = 7, 10, 13\,\mathrm{W/K}$.

further increased to 13 W/K, W reaches 50 W at $Q_H = 246$ W and efficiency is 21%. From these results, one can say that hA should be as high as possible. However, the larger heat transfer area would cause the larger viscous loss of the working gas oscillating in the heat exchanger region, which is not included in this discussion. Hence, $Q_{H,max}$ would serve as a reference input heat power that satisfies both requirements from thermal efficiency and output at a certain level when the $(hA)_{H,C}$ values are given.

Since we have mathematical expressions for W and Q_H, we are able to analytically obtain $Q_{H,max}$, and the maximum output power W_{max}. For this purpose, we seek for x that maximizes W, which is rewritten as $W = \epsilon c (hA)_H T_H f(x)$, where a function $f(x)$ is expressed as

$$f(x) = \frac{(\theta - x)(1 - x)}{\epsilon - 1 - (c + \epsilon)x}. \tag{9.50}$$

The derivative of $f(x)$ with respect to x is

$$\frac{df(x)}{dx} = \frac{(2x - \theta - 1)[\epsilon - 1 - (c + \epsilon)x] + (c + \epsilon)[x^2 - (1 + \theta)x + \theta]}{[\epsilon - 1 - (c + \epsilon)x]^2}.$$

$$(9.51)$$

By letting $df(x)/dx = 0$, we find the solutions x_1 and x_2 $(x_1 < x_2)$.

$$x_1, x_2 = \frac{\epsilon - 1 \pm \sqrt{(1 + c)[\epsilon(\theta - 1) + \theta c + 1]}}{c + \epsilon} \qquad (9.52)$$

Because $0 < \epsilon < 1$, x_1 is a negative number and only x_2 should be considered:

$$x_2 = \frac{\epsilon - 1 + \sqrt{(1 + c)[\epsilon(\theta - 1) + \theta c + 1]}}{c + \epsilon}. \qquad (9.53)$$

It should be noted that x_2 is determined mostly by $\theta = T_C/T_H$, the ratio of temperatures of the heat source and sink, because of the following reasons. Firstly, the ratio $c = (hA)_C/(hA)_H$ would be taken as around 1, if θ is not so high. Secondly, the remaining parameter ϵ is determined by the acoustic fields and thermal properties of the gas. Once x_2 is given, we have the expression for W_{max} and $Q_{H,max}$ as

$$W_{max} = \epsilon c(hA)_H T_H f(x_2) \qquad (9.54)$$

$$Q_{H,max} = (hA)_H T_H f(x_2) \qquad (9.55)$$

A rough estimate of hA may be obtained from the equations above, Eqs. (9.54) and (9.55) if the target value of the output power (heat power) is given.

9.2 Thermoacoustic Device as Nonlinear Nonequilibrium System

This textbook has discussed the thermoacoustic devices particularly when the mono-frequency oscillations of the working fluids are relatively small to warrant linear approximation. A variety of non-linear phenomena, however, is also observed. This section introduces those phenomena that are beyond the current scope of thermoacoustic theory.

9.2.1 *Shock waves, quasiperiodic oscillations, and chaos*

When a sufficiently steep temperature difference is given to a gas column of a thermoacoustic system, the gas starts to oscillate with one of the natural oscillation frequencies. The pressure waveforms are almost sinusoidal just above the critical temperature difference, but they start to distort as the amplitude is elevated with the increasing temperature difference. They finally tend to form periodic shock waves as shown in Fig. 9.7 [104]. The thermoacoustic periodic shock waves have been observed experimentally in thermoacoustic systems of the resonance tube and also of the looped tube [105], which were recently studied numerically [106].

Periodic shock waves have been observed in an acoustic resonance tube with uniform temperature when it is driven by a solid piston [107]. The essential mechanism for a shock front is the energy cascade from the driven mode to the higher harmonic oscillations through

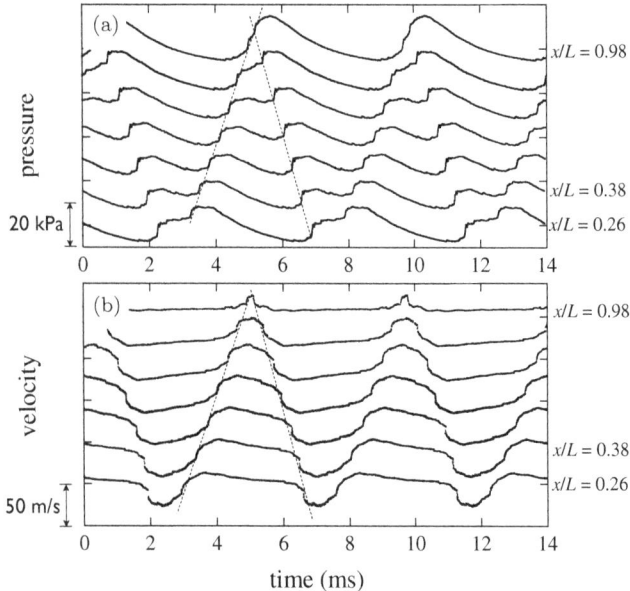

Figure 9.7: Spatio-temporal evolution of periodic shock waves induced in a resonance tube acoustic engine: (a) pressure and (b) axial velocity.

non-linear terms in the basic equations of fluid dynamics, as will be further explained. The equation of motion contains a convective term, $(u \cdot \nabla)u$, which is non-linear. When u has a time dependence of $\cos(\omega t)$, the oscillation term with angular frequency 2ω and time independent term should appear through this term as $\cos^2(\omega t) = [1 + \cos(2\omega t)]/2$. At elevated acoustic amplitudes, the higher oscillation terms are generated in the same way from the interaction between terms with ω and 2ω. Because the acoustic power supplied from the driver is distributed to a broad range of the oscillation modes, the acoustic pressure amplitude often saturates at a level of 10% of the mean pressure.

In the case of thermoacoustic shock waves, the higher modes are also created by thermoacoustic effects to create shock waves. In thermoacoustic systems, however, the overtones do not always have frequencies that are integer multiples of the fundamental frequency, because of non-uniformity of temperature. Therefore, one often encounters quasiperiodic oscillations of two or three oscillation modes, which can be easily noticed by a slow beating. In extreme temperature conditions achieved by liquid helium or combustion reaction, chaotic oscillations are generated through quasiperiodic oscillations [108, 109]. Quite recently, thermoacoustic chaotic oscillations are observed with moderate temperature conditions [110].

9.2.2 *Synchronization and amplitude death*

Self-sustained oscillations adjust the oscillation frequency when they are put under the influence of other periodic oscillations [111]. Synchronization means the frequency entrainment when the oscillation frequency coincides with the external one. It has been observed in various systems like pendulum clocks, electrical circuits, chemical reaction systems, and lasers. In acoustic systems, a system of two interacting organ pipes is a well-known example [1], which has been examined by experiments recently [112].

Synchronization has been reported in thermoacoustic systems. When a resonance tube acoustic engine is subjected to an external forcing by a loudspeaker, the oscillation frequency is pulled to the external one via saddle-node bifurcation when the external force is relatively weak, and via Hopf bifurcation when it is strong [113–115]. In addition to such a forced synchronization, a mutual synchronization is also observed in acoustic engines that are coupled by an orifice valve and/or a gas-filled narrow tube [116, 117]. Mutual interactions between two oscillators can lead to a complete annihilation of oscillations [118]. This phenomenon is called amplitude death when oscillations cease by reduction of amplitudes. As shown in Fig. 9.8, amplitude death can occur in a wide range of parameters when two acoustic engines are coupled simultaneously by using a valve and a tube.

From an experimentalist's point of view, the acoustic engines provide a good platform for synchronization studies, as the frequency and non-linear parameter are easily and precisely adjusted by the length of the tube and the temperature difference. Typical frequency of the order of hundred Hertz is not too high and not too low for conventional measurement equipments, so a long time recording of thousands of cycles and also of high time-resolution are possible. Various non-linear phenomena like synchronization of chaotic oscillations and quasiperiodic oscillations would be studied further if one chooses a thermoacoustic oscillation system as an experimental test bench.

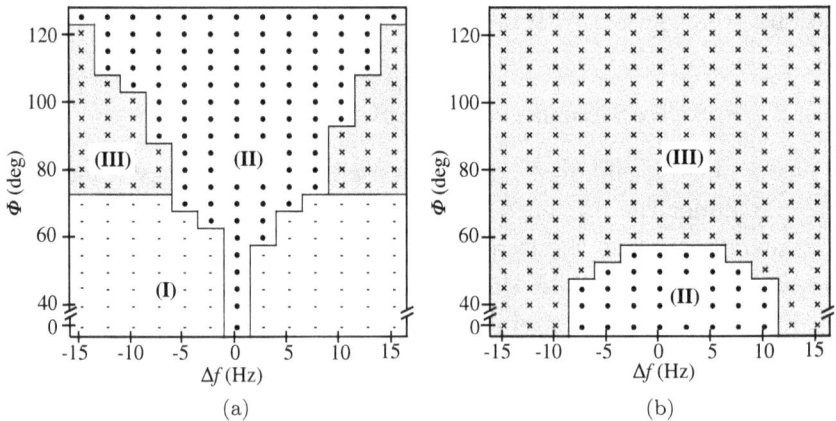

Figure 9.8: Phase diagrams when two acoustic engines are coupled: (I) asynchronous state, (II) synchronous state, (III) amplitude death [118]. The vertical axis is the coupling strength represented by the opening of the valve that connects the oscillators. The horizontal axis means the frequency detuning of the oscillators.

9.2.3 *Entropy production*

Local entropy production per unit time and volume is expressed by

$$\sigma_S = \frac{ds}{dx} \tag{9.56}$$

in a periodically oscillating thermoacoustic system, where s denotes the entropy flux density given by

$$s = \rho_m \langle\langle S'u' \rangle\rangle. \tag{9.57}$$

Tominaga has proposed entropy minimization law where the states with minimum entropy production is achieved among the states that can be induced. Consider the entropy production in a regenerator of a thermoacoustic oscillation system in Fig. 9.9. The entropy flows at ends of the regenerator are expressed by using heat flows $\tilde{Q}_{H,C}$ and temperatures $T_{H,C}$ as $\tilde{s}_{H,C} = \tilde{Q}_{H,C}/T_{H,C}$. Thus, the entropy production in the regenerator is given by $\Delta\tilde{s} = \tilde{s}_C - \tilde{s}_H$, i.e.,

$$\Delta\tilde{s} = \frac{\tilde{Q}_C}{T_C} - \frac{\tilde{Q}_H}{T_H}. \tag{9.58}$$

The first law of thermodynamics relates $\tilde{Q}_C$ and $\tilde{Q}_H$ through the increase $\Delta\tilde{I}$ of the acoustic power as $\tilde{Q}_C = \tilde{Q}_H - \Delta\tilde{I}$, and therefore, we have

$$\Delta\tilde{s} = \left(\frac{1}{T_C} - \frac{1}{T_H} \right) \tilde{Q}_H - \frac{\Delta\tilde{I}}{T_C}. \tag{9.59}$$

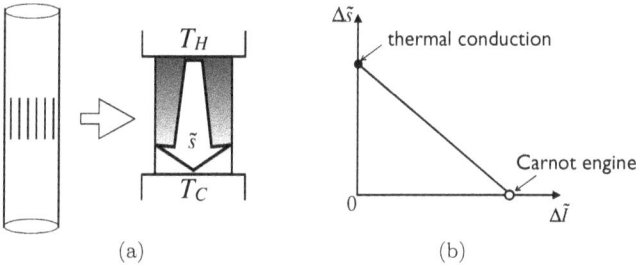

Figure 9.9: Schematic illustration of (a) a thermoacoustic oscillation system and (b) the associated entropy production.

The entropy production $\Delta \tilde{s}$ is the largest when $\Delta \tilde{I} = 0$, namely when only the thermal conduction takes place. When the oscillations take place, $\Delta \tilde{I}$ must be positive. In the Carnot engine where $\Delta \tilde{I}$ is maximized with a given temperature condition, $\Delta \tilde{s}$ that is the smallest: $\Delta \tilde{s} = 0$. In other words, thermoacoustic oscillations are induced so that the entropy production is lowered. In the same way, one can say that the entropy production is reduced when a heat pump pumps up heat from cold to hot.

Tominaga intensively discussed temperature distribution along the regenerator [119] and the thermodynamic stability of thermoacoustic oscillations in terms of entropy production minimization law [120]. He also stated that the successive mode transitions observed in thermoacoustic oscillation systems would be understood from the view point of entropy production minimization law.

It was during the industrial revolution that the second law of thermodynamics was admitted as one of the universal physical laws. The upper limit of heat engines are of primal importance for engineers that sought more efficient energy conversion devices. The relevant discussion led to the establishment of thermodynamics that we know of. Now modern society seeks a solution for energy and environmental issues. Thermoacoustic devices have presented dramatic improvement in removing moving parts from heat engines. It is interesting that the entropy production minimization law was proposed on the basis of discussion on how those thermoacoustic devices tune oscillation states without help from mechanical moving parts. Further experiments would be necessary for a better understanding of entropy production in thermoacoustic systems.

Bibliography

[1] J. W. S. Rayleigh, *The Theory of Sound*. Dover Publications, 1945.

[2] J. K. T. Feldman, "Review of the literature on Rijke thermoacoustic phenomena," *Journal of Sound and Vibration*, vol. 7, pp. 83–89, 1968.

[3] J. K. T. Feldman, "Review of the literature on Sondhauss thermoacoustic phenomena," *Journal of Sound and Vibration*, vol. 7, pp. 71–82, 1968.

[4] A. A. Putnam and W. R. Dennis, "Survey of organ-pipe oscillations in combustion systems," *Journal of Acoustical Society of America*, vol. 28, pp. 246–259, 1956.

[5] K. W. Taconis, J. J. M. Beenakker, A. O. C. Nier, and L. T. Aldrich, "Measurements concerning the vapour-liquid equilibrium of solutions of He3 in He4 below 2.19k," *Physica XV*, vol. 15, pp. 733–739, 1945.

[6] N. Rott, "Damped and thermally driven acoustic oscillations in wide and narrow tubes," *Journal of Applied Mathematics and Physics (ZAMP)*, vol. 20, pp. 230–243, 1969.

[7] N. Rott, "Thermally driven acoustic oscillations. part ii: Stability limit for helium," *Journal of Applied Mathematics and Physics (ZAMP)*, vol. 24, pp. 54–72, 1973.

[8] N. Rott, "The heating effect connected with non-linear oscillations in a resonance tube," *Journal of Applied Mathematics and Physics (ZAMP)*, vol. 25, pp. 619–634, 1974.

[9] N. Rott, "The influence of heat conduction on acoustic streaming," *Journal of Applied Mathematics and Physics (ZAMP)*, vol. 25, pp. 417–421, 1974.

[10] N. Rott, "Thermally driven acoustic oscillations, part iii: Second-order heat flux," *Journal of Applied Mathematics and Physics (ZAMP)*, vol. 26, pp. 43–49, 1975.

[11] N. Rott, "Thermally driven acoustic oscillations, part iv: Tubes with variable cross-section," *Journal of Applied Mathematics and Physics (ZAMP)*, vol. 27, pp. 197–234, 1976.

[12] G. Zouzoulas and N. Rott, "Thermally driven acoustic oscillations, part v: Gas-liquid oscillations," *Journal of Applied Mathematics and Physics (ZAMP)*, vol. 27, pp. 325–334, 1976.

[13] P. Merkli and H. Thomann, "Thermoacoustic effects in a resonance tube," *Journal of Fluid Mechanics*, vol. 70, pp. 161–177, 1975.

[14] J. C. Wheatley, "A perspective on the history and future of low-temperature refrigeration," *Physica*, vol. 109,110B, pp. 1764–1774, 1982.

[15] J. Wheatley, T. Hofler, G. W. Swift, and A. Migliori, "Understanding some simple phenomena in thermoacoustics with applications to acoustical heat engines," *American Journal of Physics*, vol. 53, pp. 147–162, February 1985.

[16] A. Tominaga, "Thermodynamic aspects of thermoacoustic theory," *Cryogenics*, vol. 35, no. 7, pp. 427–440, 1995.

[17] G. W. Swift, *Thermoacoustics*. Springer International Publishing, 2017.

[18] W. C. Ward, G. W. Swift, and J. P. Clark, "Interactive analysis, design, and teaching for thermoacoustics using DeltaEC," *Journal of Acoustical Society of America*, vol. 123, p. 3546, 2008.

[19] T. Yazaki and A. Tominaga, "Measurement of sound generation in thermoacoustic oscillations," *Proceedings of the Royal Society of London A: Mathematical, Physical and Engineering Sciences*, vol. 454, pp. 2113–2122, 1998.

[20] A. M. Fusco, W. C. Ward, and G. W. Swift, "Two-sensor power measurements in lossy ducts," *Journal of Acoustical Society of America*, vol. 91, no. 4, pp. 2229–2235, 1992.

[21] T. Biwa, Y. Tashiro, H. Nomura, Y. Ueda, and T. Yazaki, "Experimental verification of a two-sensor acoustic intensity measurement in lossy ducts," *Journal of Acoustical Society of America*, vol. 124, no. 3, pp. 1584–1590, 2008.

[22] T. Biwa, Y. Tashiro, H. Nomura, Y. Ueda, and T. Yazaki, "Acoustic intensity measurement in a narrow duct by a two-sensor method," *Review of Scientific Instruments*, vol. 78, no. 8, p. 086110, 2007.

[23] S. L. Garrett, "Thermoacoustic engines and refrigerators," *American Journal of Physics*, vol. 72, pp. 11–17, 2004.

[24] H. Ma, *Oscillating Heat Pipes*. Springer, 2015.

[25] A. A. Castrejon-Pita and G. Heulsz, "Heat-to-electricity thermoacoustic-magnetohydrodynamic conversion," *Applied Physics Letters*, vol. 90, p. 147110, 2007.

[26] T. Yoshida, T. Yazaki, H. Futaki, K. Hamguchi, and T. Biwa, "Work flux density measurements in a pulse tube engine," *Applied Physics Letters*, vol. 95, p. 0114101, 2009.

[27] T. Yazaki, A. Iwata, T. Maekawa, and A. Tominaga, "Traveling wave thermoacoustic engine in a looped tube," *Physical Review Letters*, vol. 81, pp. 3128–3131, 1998.

[28] S. Backhaus and G. W. Swift, "A thermoacoustic stirling heat engine," *Nature*, vol. 399, pp. 335–338, 1999.

[29] H. Tijani and S. Spoelstra, "A high performance thermoacoustic engine," *Journal of Applied Physics*, vol. 110, p. 093519, 2011.

[30] C. D. West, *Liquid Piston Stirling Engines*. Van Nostrand Reinhold Company, 1983.

[31] P. Ceperley, "A pistonless stirling engine—the traveling wave heat engine," *Journal of Acoustical Society of America*, vol. 66, pp. 1508–1513, 1979.

[32] J. C. Wheatley, "Intrinsically irreversible or natural engines," in *Frontiers in Physical Acoustics, Proceedings of the E. Fermi Summer School*, pp. 395–475, 1986.

[33] G. W. Swift, "Thermoacoustic engines and refrigerators," *Physics Today*, vol. 48, pp. 22–28, 1995.

[34] G. Swift, D. Gardner, and S. Backhauss, "Acoustic recovery of lost power in pulse tube refrigerators," *Journal of Acoustical Society of America*, vol. 105, pp. 711–724, 1999.

[35] H. Tijani and S. Spoelstra, "Study of a coaxial thermoacoustic-stirling cooler," *Cryogenics*, vol. 48, pp. 77–82, 2008.

[36] E. I. Mikulin, A. A. Tarasov, and M. P. Shkrebyonck, "Low-temperature expansion pulse tubes," *Advances in Cryogenic Engineering*, vol. 29, pp. 629–637, 1984.

[37] K. Kanao, N. Watanabe, and Y. Kanazawa, "A miniature pulse tube refrigerator for temperatures below 100 K," *Cryogenics*, vol. 34, pp. 167–170, 1984.

[38] S. Zhu, P. Wu, and Z. Chen, "Double inlet pulse tube refrigerators: An important improvement," *Cryogenics*, vol. 30, pp. 514–520, 1990.

[39] T. Yazaki, T. Biwa, and A. Tominaga, "A pistonless Stirling cooler," *Applied Physics Letters*, vol. 80, no. 1, pp. 157–159, 2002.

[40] S. Hasegawa, T. Yamaguchi, and Y. Oshinoya, "A thermoacoustic refrigerator driven by a low temperature-differential, high-efficiency multistage thermoacoustic engine," *Applied Thermal Engineering*, vol. 58, no. 1, pp. 394–399, 2013.

[41] U. H. Kurzweg and L. Zhao, "Heat transfer by high frequency oscillations: A new hydrodynamic technique for achieving large

effective thermal conductivities," *The Physics of Fluids*, vol. 27, pp. 2624–2627, 1984.

[42] M. Ozawa, T. Sakaguchi, H. Hamaguchi, A. Kawamoto, A. Ichii, and S.Ono, "Enhancement of heat transfer by sinusoidal oscillation of fluid (transient behavior of a dream pipe)," *Transactions of the Japan Society of Mechanical Engineers, Series B*, vol. 56, no. 530, pp. 3056–3063, 1990 (in Japanese).

[43] T. B. Gabrielson, "Background and perspective: William Derham's de motu soni (on the motion of sound)," *Acoustics Today*, pp. 17–26, January 2009.

[44] H. Tijdeman, "On the propagation of sound waves in cylindrical tubes," *Journal of Sound and Vibration*, vol. 39, no. 1, pp. 1–33, 1975.

[45] T. Yazaki, Y. Tashiro, and T. Biwa, "Measurements of sound propagation in narrow tubes," *Proceedings of the Royal Society of London A: Mathematical, Physical and Engineering Sciences*, vol. 463, no. 2087, pp. 2855–2862, 2007.

[46] M. Ohmi and M. Iguchi, "Critical Reynolds number in an oscillating pipe flow," *Bulletin of JSME*, vol. 25, no. 200, pp. 165–172, 1982.

[47] M. Ohmi, M. Iguchi, K. Kakehashi, and T. Masuda, "Transition to turbulence and velocity distribution in an oscillating pipe flow," *Bulletin of JSME*, vol. 25, no. 201, pp. 365–371, 1982.

[48] M. Ohmi, M. Iguchi, and I. Urahata, "Flow patterns and frictional losses in an oscillating pipe flow," *Bulletin of JSME*, vol. 25, no. 202, pp. 536–543, 1982.

[49] M. Iguchi, M. Ohmi, and K. Maegawa, "Analysis of free oscillating flow in a U-shaped tube," *Bulletin of JSME*, vol. 25, no. 207, pp. 1398–1405, 1982.

[50] G. W. Swift, "Thermoacoustic engine," *Journal of Acoustical Society of America*, vol. 84, pp. 1145–1180, 1988.

[51] T. Biwa, Y. Ueda, H. Nomura, U. Mizutani, and T. Yazaki, "Measurement of the Q value of an acoustic resonator," *Physical Review E*, vol. 72, p. 026601, August 2005.

[52] H. A. Kramers, "Vibrations of a gas column," *Physica XV*, pp. 971–984, December 1949.

[53] T. Yazaki, A. Tominaga, and Y. Narahara, "Experiments on thermally driven acoustic oscillations of gaseous helium," *Journal of Low Temperature Physics*, vol. 41, pp. 45–60, Oct 1980.

[54] A. A. Atchley, H. E. Bass, T. J. Hofler, and H. T. Lin, "Study of a thermoacoustic prime mover below onset of self-oscillation," *Journal of Acoustical Society of America*, vol. 91, pp. 734–743, 1992.

[55] A. A. Atchley, "Standing wave analysis of a thermoacoustic prime mover below onset of self-oscillation," *Journal of Acoustical Society of America*, vol. 92, pp. 2907–2914, 1992.

[56] A. A. Atchley, "Analysis of the initial buildup of oscillations in a thermoacoustic prime mover," *Journal of Acoustical Society of America*, vol. 95, pp. 1661–1664, 1994.

[57] I. Urieli and D. M. Berchowitz, *Stirling Cycle Engine Analysis*. Bristol: Adam Hilger Ltd., 1984.

[58] A. Widyaparaga, T. Hiromatsu, Deendarlianto, M. Kohno, and Y. Takata, "Acoustic field alteration in a 100 Hz dual acoustic driver straight tube travelling wave thermoacoustic heat pump for thermoacoustic heat transport control," *International Journal of Heat and Mass Transfer*, vol. 151, p. 119274, 2020.

[59] Y. Ueda and T. Biwa, "Efficiencies of a pulse-tube refrigerator and heat-driven thermoacoustic cooler," *TEION KOGAKU*, vol. 41, no. 2, pp. 73–80, 2006 (in Japanese).

[60] D. L. Gardner and G. W. Swift, "A cascade thermoacoustic engine," *Journal of Acoustical Society of America*, vol. 114, no. 4, pp. 1905–1919, 2003.

[61] T. Biwa and K. Takao, "Acoustic power amplification by multiple regenerators," *TEION KOGAKU*, vol. 47, no. 1, pp. 42–46, 2012 (in Japanese).

[62] T. Biwa, D. Hasegawa, and T. Yazaki, "Low temperature differential thermoacoustic Stirling engine," *Applied Physics Letters*, vol. 97, no. 3, p. 034102, 2010.

[63] K. de Blok, "Novel 4-stage traveling wave thermoacoustic power generator," in *Proceedings of ASME 2010 3rd Joint US-European Fluids Engineering Summer Meeting and 8th International Conference on Nanochannels, Microchannels and Minichannels*, (Montreal, Canada), 73–79, https://asmedigitalcollection.asme.org/FEDSM/proceedings-abstract/FEDSM2010/49491/73/349639, August 2010.

[64] T. Inoue and S. Kawano, "The diagrammatical understanding of thermoacoustic theory," *TEION KOGAKU*, vol. 29, no. 11, pp. 558–567, 1994 (in Japanese).

[65] P. H. Ceperley, "Gain and efficiency of a traveling wave heat engine," *Journal of Acoustical Society of America*, vol. 72, pp. 1688–1694, 1982.

[66] P. H. Ceperley, "Gain and efficiency of a short traveling wave heat engine," *Journal of Acoustical Society of America*, vol. 77, pp. 1239–1244, 1985.

[67] T. Biwa, R. Komatsu, and T. Yazaki, "Acoustical power amplification and damping by temperature gradients," *Journal of Acoustical Society of America*, vol. 129, pp. 132–137, 2011.

[68] Y. Ueda and C. Kato, "Stability analysis of thermally induced spontaneous gas oscillations in straight and looped tubes," *Journal of Acoustical Society of America*, vol. 124, pp. 851–858, 2008.

[69] Y. Ueda, T. Biwa, U. Mizutani, and T. Yazaki, "Acoustic field in a thermoacoustic Stirling engine having a looped tube and resonator," *Applied Physics Letters*, vol. 81, pp. 5252–5254, 2002.

[70] Y. Ueda, T. Biwa, U. Mizutani, and T. Yazaki, "Experimental studies of a thermoacoustic Stirling prime mover and its application to a cooler," *Journal of Acoustical Society of America*, vol. 115, pp. 1134–1141, 2004.

[71] S. Backhaus, E. Tward, and M. Petach, "Traveling-wave thermoacoustic electric generator," *Applied Physics Letters*, vol. 85, pp. 1085–1087, 2004.

[72] E. Luo, W. Dai, Y. Zhang, and H. Ling, "Thermoacoustically driven refrigerator with double thermoacoustic-Stirling cycles," *Applied Physics Letters*, vol. 88, p. 074102, 2006.

[73] H. Hatori, T. Biwa, and T. Yazaki, "How to build a loaded thermoacoustic engine," *Journal of Applied Physics*, vol. 111, p. 074905, 2012.

[74] V. Zorgonotti, G. Pelenet, G. Poignand, and S. L. Garrett, "Prediction of limit cycle amplitudes in thermoacoustic engines by means of impedance measurements," *Journal of Applied Physics*, vol. 124, p. 154901, 2018.

[75] W. R. Martini, "Test on a 4U tube heat operated heat pump," in *Proceedings of the Intersociety Energy Conversion Conference*, 872–874, 1983.

[76] D. H. Li, Y. Chen, E. C. Luo, and Z. Wu, "Study of a liquid-piston traveling-wave thermoacoustic engine with different working gases," *Energy*, vol. 74, pp. 158–163, 2014.

[77] H. Hyodo, S. Tamura, and T. Biwa, "A looped-tube traveling-wave engine with liquid pistons," *Journal of Applied Physics*, vol. 122, p. 114902, 2017.

[78] S. Tamura, H. Hyodo, and T. Biwa, "Experimental and numerical analysis of a liquid-piston Stirling engine with multiple unit sections," *Japanese Journal of Applied Physics*, vol. 58, p. 017001, 2019.

[79] J. Wheatley, T. Hofler, G. W. Swift, and A. Migliori, "An intrinsically irreversible thermoacoustic heat engine," *Journal of Acoustical Society of America*, vol. 74, pp. 153–170, 1983.

[80] J. Wheatley and A. Cox, "Natural engines," *Physics Today*, vol. 38, pp. 50–58, 1985.

[81] S. L. Garrett, J. A. Adeff, and T. J. Hofler, "Thermoacoustic refrigerator for space applications," *Journal of Thermophysics and Heat Transfer*, vol. 7, pp. 595–599, 1993.

[82] Y. Ueda, B. M. Mehdi, K. Tsuji, and A. Akisawa, "Optimization of the regenerator of a traveling-wave thermoacoustic refrigerator," *Journal of Applied Physics*, vol. 107, p. 034901, 2010.

[83] M. E. Poese, R. W. M. Smith, and S. L. Garrett, "Regenerator based thermoacoustic refrigerator for ice cream storage applications," *Journal of Acoustical Society of America*, vol. 114, no. 4, pp. 2328–2328, 2003.

[84] M. Sato, S. Hasegawa, T. Yamaguchi, and Y.Oshinoya, "Experimental evaluation of performance of double-loop thermoacoustic refrigerator driven by multistage thermoacoustic engines," in *Proceedings of ICEC 24-ICMC 2012*, pp. 391–394, 2013.

[85] S. Sunahara, T. Biwa, and U. Mizutani, "Thermoacoustic heat pumping effect in a Gifford-McMahon refrigerator," *Journal of Applied Physics*, vol. 92, pp. 6334–6336, 2002.

[86] R. D. McCarty, *Thermophysical Properties of Helium-4 from 2 to 1500 K with Pressures to 1000 Atmospheres*. National Bureau of Standards, 1972.

[87] T. Iwase, T. Biwa, and T. Yazaki, "Acoustic impedance measurements of pulse tube refrigerators," *Journal of Applied Physics*, vol. 107, p. 034903, 2010.

[88] K. Nakamura and Y. Ueda, "Design and construction of a standing-wave thermoacoustic engine with heat sources having a given temperature ratio," *Journal of Thermal Science and Technology*, vol. 6, pp. 416–423, 2011.

[89] H. Hyodo, M. Muraoka, and T. Biwa, "Stability analysis of thermoacoustic gas oscillations through temperature ratio dependence of the complex frequency," *Journal of Physical Society of Japan*, vol. 86, p. 104401, 2017.

[90] J. R. Olson and G. W. Swift, "Energy dissipation in oscillating flow through straight and coiled pipes," *Journal of Acoustical Society of America*, vol. 100, pp. 2123–2131, 1996.

[91] B. Smith and G. W. Swift, "Power dissipation and time-averaged pressure in oscillating flow through a sudden area change," *Journal of Acoustical Society of America*, vol. 113, pp. 2455–2563, 2003.

[92] A. Petculescu and L. A. Wilen, "Oscillatory flow in jet pumps: Nonlinear effects and minor losses," *Journal of Acoustical Society of America*, vol. 113, pp. 1282–1292, 2003.

[93] H. Bailliet, V. Gusev, R. Raspet, and R. Hiller, "Acoustic streaming in closed thermoacoustic devices," *Journal of Acoustical Society of America*, vol. 110, pp. 1808–1821, 2001.

[94] J. R. Olson and G. W. Swift, "Acoustic streaming in pulse tube refrigerators: Tapered pulse tubes," *Cryogenics*, vol. 37, pp. 769–776, 1997.

[95] Y. Ueda, S. Yonemitsu, K. Ohashi, and T. Okamoto, "Measurement and empirical evaluation of acoustic loss in tube with abrupt area change," *Journal of Acoustical Society of America*, vol. 147, no. 1, pp. 364–370, 2020.

[96] Y. Ueda, T. Kato, and C. Kato, "Experimental evaluation of the acoustic properties of stacked-screen regenerators," *Journal of Acoustical Society of America*, vol. 125, no. 2, pp. 780–786, 2009.

[97] S. H. Hsu and T. Biwa, "Modeling of a stacked-screen regenerator in an oscillatory flow," *Japanese Journal of Applied Physics*, vol. 56, no. 1, p. 017301, 2016.

[98] A. Piccolo and G. Pistone, "Estimation of heat transfer coefficients in oscillating flows: The thermoacoustic case," *International Journal of Heat and Mass Transfer*, vol. 49, pp. 1631–1642, 2006.

[99] J. A. de Jong, Y. H. Wijnant, and A. de Boer, "A one-dimensional heat transfer model for parallel-plate thermoacoustic heat exchangers," *Journal of Acoustical Society of America*, vol. 135, pp. 1149–1158, 2014.

[100] G. W. Swift, "Analysis and performance of a large thermoacoustic engine," *Journal of Acoustical Society of America*, vol. 92, pp. 1551–1563, 1992.

[101] G. Mozurkewich, "A model for transverse heat transfer in thermoacoustics," *Journal of Acoustical Society of America*, vol. 103, pp. 3318–3326, 1998.

[102] G. Mozurkewich, "Heat transfer from transverse tubes adjacent to a thermoacoustic stack," *Journal of Acoustical Society of America*, vol. 110, pp. 841–847, 2001.

[103] T. Biwa, T. Kobayashi, and H. Hyodo, "Direct measurements of steady heat transfer maintained by oscillating pipe flow in thermoacoustic system," *Journal of Applied Physics*, vol. 125, p. 014903, 2019.

[104] T. Biwa, K. Sobata, and T. Yazaki, "Observation of thermoacoustic shock waves in a resonance tube (L)," *Journal of Acoustical Society of America*, vol. 136, pp. 965–968, 2014.

[105] T. Biwa, T. Takahashi, and T. Yazaki, "Observation of traveling wave thermoacoustic shock waves," *Journal of Acoustical Society of America*, vol. 130, pp. 3558–3561, 2011.

[106] C. Olivier, G. Penelet, G. Poignand, J. Gilbert, and P. Lotton, "Weakly nonlinear propagation in thermoacoustic engines: A numerical study of higher harmonics generation up to the appearance of shock waves," *Acta Acustica United with Acustica*, vol. 101, pp. 941–949, 2015.

[107] D. B. Cruikshank, Jr., "Experimental investigation of finite-amplitude acoustic oscillations in a closed tube," *Journal of Acoustical Society of America*, vol. 52, pp. 1024–1036, 1972.

[108] T. Yazaki, "Experimental observation of thermoacoustic turbulence and universal properties at the quasiperiodic transition to chaos," *Physical Review E*, vol. 48, pp. 1806–1818, 1993.

[109] L. Kabiraji, A. Saaurabh, P. Wahi, and R. I. Sujith, "Route to chaos for combustion instability in ducted laminar premixed flames," *Chaos*, vol. 22, p. 023129, 2012.

[110] R. Delage, Y. Takayama, and T. Biwa, "On-off intermittency in coupled chaotic thermoacoustic oscillations," *Chaos*, vol. 27, p. 043111, 2017.

[111] A. Pikovsky, M. Rosenblum, and J. Kurth, *Synchronization: A Universal Concept in Nonlinear Sciences*. Cambridge University Press, 2001.

[112] M. Abel, S. Bergweiler, and R. Gerhard-Multhaupt, "Synchronization of organ pipes: Experimental observations and modeling," *Journal of Acoustical Society of America*, vol. 119, pp. 2467–2475, 2006.

[113] T. Yoshida, T. Yazaki, Y. Ueda, and T. Biwa, "Forced synchronization of periodic oscillations in a gas column: Where is the power source?," *Journal of Physical Society of Japan*, vol. 82, p. 103001, 2013.

[114] G. Penelet and T. Biwa, "Synchronization of a thermoacoustic oscillator by an external sound source," *American Journal of Physics*, vol. 81, pp. 290–297, 2013.

[115] H. Hyodo and T. Biwa, "Phase-locking and suppression states observed in forced synchronization of thermoacoustic oscillator," *Journal of Physical Society of Japan*, vol. 87, p. 034402, 2018.

[116] P. S. Spoor and G. W. Swift, "Mode locking of acoustic resonators and its application to vibration cancellation in acoustic heat engines," *Journal of Acoustical Society of America*, vol. 106, pp. 1353–1362, 1999.

[117] P. S. Spoor and G. W. Swift, "The Huygens entrainment phenomenon and thermoacoustic engines," *Journal of Acoustical Society of America*, vol. 108, pp. 588–599, 2000.

[118] T. Biwa, S. Tozuka, and T. Yazaki, "Amplitude death in coupled thermoacoustic oscillators," *Physical Review Applied*, vol. 3, p. 034006, 2015.

[119] A. Tominaga, "Stable temperature gradients of a short regenerator induced by forced fluid oscillations," *Journal of Cryogenics and Superconductivity Society of Japan*, vol. 39, pp. 632–637, 2004.

[120] A. Tominaga, "Simple heat-conduction and thermoacoustic self-sustained oscillation branches," *Journal of Cryogenics and Superconductivity Society of Japan*, vol. 40, pp. 13–21, 2005.

[121] P. Murti, A. Takizawa, E. Shoji, and T. Biwa, "Design guideline for multi-cylinder-type liquid-piston Stirling engine", submitted to Applied Thermal Engineering.

Index

www.ingramcontent.com/pod-product-compliance
Lightning Source LLC
Chambersburg PA
CBHW050540190326
41458CB00007B/1850